THE BEST THINGS IN LIFE ARE FREE

ARTS & CULTURE

OUTDOORS & ADVENTURE

WELLNESS

FOOD & DRINK

CONTENTS

CONTENTS

INTRODUCTION

When we took our children on their first foray into Africa we visited Marrakesh in Morocco. On our first evening we walked into the Djemaa el-Fna (p16), the city's world famous open square. It was like walking into another world. The smoke and smells from food stalls, the sound of drummers and the clamour of people shouting, singing and hawking their wares. Everyone's eyes were out on stalks, senses in overdrive. I had been worried it would be too much, but the kids asked to go again the following night. One of the most memorable moments I've had travelling cost nothing beyond the gumption to walk into the square that night.

The monetary value implied in the term *free* can misrepresent what's on offer within these pages. The quality of an experience, after all, is not attached to a price tag. Many of the suggestions here involve unearthing the world's secret wonders, whether that's swimming the idyllic shores of Uoleva, Tonga (p308) or strolling the tombs and monuments of Delhi's Lodi Gardens (p54). Walkers tackling the great tracks of New Zealand (p300) will find themselves close to the soul of those beautiful islands.

For many of us, when we take our first steps travelling free things are not only appealing but essential if we're to make our backpacking days last as long as possible. And it's not just formative forays – many unforgettable blasts of freedom and discovery tend to be budget affairs. You quickly realise that cheap can mean much, much better. No Roman dinner will ever match the bread-and-cheese picnic in Villa Celimontana, a short walk from the Colosseum, on my first visit to the Eternal City. And if we're talking life lessons, there are few better insights into the human condition than sharing a dorm room with a dozen others from around the world.

Nations all around the world recognise the value of making the wonders under their stewardship accessible. 5000 years of Chinese history? Free (National Museum of China, p59). 19 museums and galleries in Washington DC? Free (Smithsonian Institution, p174). Britain's national parks? All free. Donations always welcome, of course. And if you want to experience all those popular places at their best then get up early and get their before the crowds arrive. The priceless calm of early mornings doesn't come with a price tag either, by the way.

It's an exaggeration to say that everything good is free, so you'll find plenty of excellent value cheap things to experience throughout this book. Dip into your spare change for classy street food like a *choripán* (chorizo sandwich) in Buenos Aires (p234), the top temples to visit in Bangkok, Thailand (p56) and Northern California's best farmers' markets (p186). Great memories, a happier you and a grateful wallet; one glance through these pages and you may never aspire to the indulgences of top-end travel again.

Tom Hall

INTRODUCTION

THE BEST PLACES FOR YOU

Borneo, *p86*

Serengeti and Masai Mara, *p38*

Reykjavík & Southern Iceland, *p160*

Bali, Lombok & the Gili Islands, *p92*

Atlas Mountains, *p34*

Helsinki, *p163*

The Caribbean, *p216*

Central Italy (Florence & Tuscany, Rome), *p130*, *p143*

Budapest, *p162*

Chiang Mai Region, *p70*

Goa & Kerala, *p100*

The Cape, *p22*

Polynesia, *p308*

Mauritius & Reunion, *p45*

Stockholm & the Archipelago, *p161*

Istanbul, *p117*

Copenhagen, *p135*

Above: bathing in Budapest.

Right: streetside snacking in Singapore.

Hong Kong, *p66*

Singapore, *p67*

INTRODUCTION

Above: find affordable safaris in Africa.

Below: city sights in London.

Amazon, *p250*

Patagonia, *p254*

Egypt (Cairo, Red Sea Region), *p14, p36*

Vancouver & British Columbia, *p202*

The Ruta Maya, *p206*

Wellington & North Island (with Auckland), *p272, p282, p300*

California (NorCal, California, Los Angeles), *p178, p186, p214*

Venice & Alpine Italy & Slovenia, *p116, p148*

London, *p112*

Delhi, *p54*

Queenstown & South Island, Sydney & New South Wales, *p260, p280, p296, p309*

Washington DC, *p174*

Paris, *p114*

Moscow, *p120*

Bangkok, *p56*

Buenos Aires, *p226*

New York City, *p176*

San Sebastián, *p138*

Toronto, *p191*

● WELLNESS

● FOOD & DRINK

● ARTS & CULTURE

● OUTDOORS & ADVENTURE

INTRODUCTION

THE BEST THINGS IN LIFE IN AFRICA: TOOLKIT

Africa can be as cheap as nyama choma (grilled meat) or as expensive as diamonds, depending on what you do and where you go. Pick cheaper national parks, stay in tented camps over lodges, and take public transport, and you will go further for less.

In theory, Africa should be one of the least expensive destinations on the planet. Food and transport are cheap, camping sites are everywhere, and Africa's biggest tourist attraction – its amazing wildlife – wanders into the frame for free. In reality, however, Africa can suck notes from your wallet faster than Tokyo or New York.

To keep a lid on costs, aim to camp, eat at local restaurants, travel cheaply on local buses and trucks, and take your safaris in less famous national parks – you'll pay less and have a more interesting trip than the millionaires buzzing in and out by chartered plane.

TOP BUDGET DESTINATIONS IN AFRICA

Travelling in Africa can eat through money like lions at a kill; make your funds go further in the following bargain locations:

• **CAPE TOWN:** A thriving backpacker scene pulls down prices in South Africa's Mother City.
• **CAIRO:** Balance the expensive (the Pyramids et al) with abundant free sights.
• **MOROCCO:** Much to see for free if you skip expensive *riad* hotels.

• **ETHIOPIA:** Good food and lots of history for not a lot of bucks.
• **MALAWI:** Cheap access to wildlife and a refreshingly laid-back mood.

TRANSPORT

Don't underestimate the challenges of long-distance travel in Africa. Flat, smooth tarmac becomes a memory once you leave the coast, and public transport quickly deteriorates into a rag-

Below: safaris need not cost a fortune if you chose your country carefully.

© David Malan | Getty Images

tag collection of worn-out buses, freight trucks and shared taxis. On the flip side, if you have the time to travel as locals do, you'll rarely have to dig deep to get from A(ccra) to B(ulawayo).

AIR
Domestic flights are often great value, particularly as an alternative to the gruelling overland journey across the Sahara. Safety standards can be a worry, but this has to be balanced against the risks of travel by land in insecure regions. South African Airways (flysaa.com), Royal Air Maroc (royalairmaroc.com), EgyptAir (egyptair.com), and Kenya Airways (kenya-airways.com) have better reputations than most. Check the safety record of all carriers at airlineratings.com.

Airpasses offer decent savings when booked with an inbound flight. Star Alliance's Africa Airpass (staralliance.com) covers 30 airports in 23 countries served by South African Airways, Ethiopian Airlines and EgyptAir. OneWorld's Visit Africa Pass covers nine southern African destinations served by British Airways, Comair and other partner airlines (oneworld.com). North Africa is covered by some European airpasses (p105).

• **Budget tips** – Africa's budget airlines offer cheap seats, but hold bags are always extra; try Kulula (kulula.com), Fastjet (fastjet.com), Mango (flymango. com) and Fly540 (fly540.com).

• **Travel light** – you'll save on baggage costs on budget airlines, and many safari flights use small aircraft with even smaller baggage allowances (15kg or less).

TRAIN
Railways played a vital role in trade during colonial times, but the networks have atrophied. South Africa and Namibia have the most useful rail services, but

there are also some rewarding journey-is-the-destination routes in Kenya, Mali, Gabon, Senegal and Mauritania. Fares are comparable to buses, or slightly more for sleeping carriages.

BUSES, TRUCKS AND MATATUS
Buses are the backbone of rural travel in Africa, though some vehicles only just warrant the description – hard seats, clunky suspension and curtainless windows that let in dust and the sun's heat all add to the fun.

• **Master the minibus** – as well as full-sized buses, many trips can be covered cheaply by minibus (they're known as *matatus* in East Africa, and by a host of other names elsewhere) for a not much higher fare.

• **Go by bush taxi** – ageing Peugeot station wagons and estate cars function as shared taxis across Africa. They're rarely expensive, often overloaded, and go everywhere.

• **Take the truck** – freight trucks provide

Above: the luxury Blue Train travels between Cape Town and Pretoria in South Africa but it's not a bargain option.

AFRICA

Above: camping, here in Namibia, is a low-cost accommodation option in rural areas.

an informal bus service in remote areas; you'll most likely be dumped in the back on the top of the load, but you'll pay less than on the bus.

• **Ride in comfort** – in the far north and far south of Africa, comfortable, air-con buses offer reclining seats for a not-too-painful-price; you might even get a few hours of sleep.

BICYCLE & MOTORCYCLE

Hundreds of adventurous travellers cross the continent on motorcycles and mountain bikes (better than touring bikes on unsealed roads). The ferry from Europe to Morocco provides an easy start to the journey; carry plenty of water, parts and tools, and camping gear in case you get caught outdoors overnight.

CAR & MOTORCYCLE

Car-rental in Africa can be a daunting experience given the often awful local driving conditions, but a self-drive 4WD is a great way to reach many national parks. Rates for 4WDs start from US$150 a day, so get a group together and split the costs.

• **2WD or 4WD?** – in North Africa and Southern Africa, you can get around comfortably in a conventional car, including inside many national parks.

• **Take a driver** – in some countries, renting a car with a driver (who knows the local road rules) doesn't cost much more than renting just the car.

• **Watch the excess** – it can be US$2000 or more; take extra insurance to avoid a painful shock in the event of an accident.

BOAT

Boats of all sizes and configurations traverse Africa's lakes and rivers, offering an atmospheric way to cover long distances. Travel by water can be dirt cheap (if you travel on deck with the cargo) or painfully expensive (if you require a cabin, air-con and access to a dining room). Be wary of boarding overloaded boats – accidents are not uncommon.

AFRICA

LOCAL TRANSPORT

Generally, you'll be piling into cramped and overcrowded but pocket-friendly buses, minibuses and pickup trucks, or taking a taxi (shared or otherwise). Indian-style autorickshaws are starting to pop up in many countries.

- **Minibus safety** – low-cost minibuses buzz around like bees in most African cities, but head-on collisions are common so avoid the so-called 'death seat' next to the driver.
- **Rideshare** – Uber is making ground in Africa; useful local apps include South Africa's Jrney (jrney.co) and Bolt (bolt.eu). Most are integrated with mobile phone-based cashless payment systems such as M-Pesa and Paga.

ACCOMMODATION

Quality hostel accommodation can be found easily in North Africa and Southern Africa but elsewhere, cheap digs can be alarmingly run-down, and even dangerous for solo women travellers. Many travellers are happy to pay higher prices for more comfortable accommodation in safer parts of town.

BUDGET SLEEPING

- **Sleep under canvas** – low-cost camping is the default accommodation option in many rural areas, whether in designated campsites (including in national parks) or the gardens of willing hostels and hotels.
- **Tented camps** – on safari, tented camps (usually safari tents with their own bathroom) are cheaper than park lodges, particularly if you stay in buffer zones just outside national parks.
- **Stay with locals** – when off the beaten track, space for a tent or sleeping bag can often be negotiated with villagers for a modest fee.

- **Hostel tips** – hostels dot the main routes used by overland travellers (Nairobi to Cape Town, Cairo to Nairobi etc), offering dorms and box rooms (and often camping), plus discounted tours and safaris. Hostelworld (hostelworld.com) has plenty of listings.

HOTELS, B&BS & GUESTHOUSES

- **B&B basics** – the terms B&B and guesthouse are used interchangeably; they range from pocket-pleasing rooms in family homes to lavish colonial-era mansions. In Francophone West Africa, look out for chambres d'hôtes or maisons d'hôtes.
- **Airbnb Africa** – the independent bed-booking site has lots of cheap rooms in South Africa, Kenya, Nigeria, Ghana, Rwanda and Mozambique.
- **Cheap hotels 101** – in Africa's cheapest hotels, shared bathrooms are the norm, insect-life can be plentiful and air-con and mosquito nets non-existent. Some cheap hotels are fronts for prostitution; solo women travellers should be cautious about staying in bare-bones places.
- **Learn the lexicon** – budget hotels are known variously as rest houses, *pensao* (in Mozambique), *campement* (in West Africa) and *gesti* or lodgings in East Africa. The term 'hotel' or *hoteli* often refers to a basic eating place, not somewhere to stay.

PARK LODGES

Lodges inside national parks range from the atmospheric to the extraordinary, but because of the drawcard of wildlife drinking at the waterhole in front of the dining room, the sky's the limit when it comes to prices. You'll often find cheaper lodge-style digs just outside the park boundaries.

PRO TIPS

In my many years of travel across dozens of African nations I've found that vast sums of money can happily disappear in the blink of an eye – a US$1500 mountain gorilla permit, a few nights in an Okavango Delta lodge, a flight safari over the Skeleton Coast. Yet I've also made paltry sums last weeks on end. Non-touristy lodgings and food are always budget friendly, as is transportation – I've found bargain rides on the back of transport trucks, squeezed into Sahelian bush taxis, taken countless public minibuses and buses, and even snagged a spare seat in a Land Rover travelling from Chad to Morocco.

Matt Phillips,
Lonely Planet
writer, editor

AFRICA

ARTS & CULTURE

Africa is less about Old Masters at big-name museums and more about big cultural encounters in no-name locations, but the continent has its share of don't-miss institutions, particularly in North Africa and in South Africa's main city hubs.

More often, though, African culture is a living thing, viewed not in museums but in rural village markets and at tribal ceremonies. Don't expect the spectacle to be completely free: locals expect tourists to contribute to the economy, be that via tips for musicians and dancers or a donation to the people you take photos of.

Above: Hatchepsut Temple in the Valley of the Nobles in Luxor, Egypt. Right: the Koutoubia Mosque, the largest in Marrakesh.

AFRICA'S TOP SWITCHEROOS

Africa's top sights are world-class, but that often translates into a weighty entry fee; try some of the following swaps to save.

- **SKIP:** Valley of the Kings, Luxor, Egypt (entry US$10.20)
- **SEE:** Tombs of the Nobles, Luxor, Egypt (entry from US$1.27)
- **SKIP:** Robben Island, Cape Town, South Africa (US$22.75)
- **SEE:** District Six Museum, Cape Town, South Africa (entry US$2.68)
- **SKIP:** El Jem Amphitheatre, El Djem, Tunisia (entry US$3.70)
- **SEE:** Tipasa Archaeological Park, Tipaza, Algeria (entry US$0.61)
- **SKIP:** Cairo Tower, Cairo, Egypt (entry US$4.46)
- **SEE:** Minaret of the Mosque of Ibn Tulun, Cairo, Egypt (free)
- **SKIP:** Ruins of Great Zimbabwe, Zimbabwe (entry US$15)
- **SEE:** Ruins of Aksum, Ethiopia (entry from US$1.30)

AFRICA'S CULTURE FOR LESS

Africa is less about free, and more about affordable. Most museums, galleries and

sights charge, but entry fees are rarely going to leave a gaping hole in your travel funds, except at Egypt's top ancient sites. Engaging with Africa's amazing tribal culture on a budget is harder; encounters with the continent's diverse tribal peoples typically involve money changing hands, even for taking a photo.

Fixed-price, organised cultural events can provide a less commercial-feeling experience than the setting might suggest, without the awkwardness of requests for tips. One great (and free) way to engage with African culture on an even footing is in church, where devotional songs fill the air with passionate fervour.

- **Occasional student discounts** – South Africa has decent discounts for student card holders, and there are some meaty savings in North Africa for holders of an ISIC (isic.org) card, including at Egypt's top museums.
- **Make for the markets** – even in big cities, markets lure in tribal people from the countryside, offering glimpses of Africa's complex cultures without the cost of tribal tours.
- **Have faith** – many mosques, churches and other sacred sites can be visited, respectfully, for free, and locals are often keen to tell the story of their customs and beliefs; ask before entering so as not to cause offense.
- **Lost civilisations** – there's more to see than the pyramids. Take in desert forts and Roman ruins in North Africa, toppled Swahili trading posts in East Africa and the vast walls of Great Zimbabwe. If you exclude Egypt's top-tier temples and funerary complexes, entry fees won't stretch your budget.
- **Get a pass** – the Cape Town pass (capetownpass.com) offers free access to more than 30 sights for R645/945 (two/three days).

HIGHLIGHTS

KHAN AL KHALILI, CAIRO
The Egyptian Museum powerfully (and rather expensively) recreates Egypt's past, but some of that past lives on in the tangled bazaars of Cairo's ancient Khan Al Khalili market place. It's sometimes touristy, sometimes authentic, and always atmospheric. *p14*

DISTRICT SIX, CAPE TOWN
The museum at the heart of the apartheid-ravaged community of District Six tells a powerful story of oppression, an injustice that director Neill Blomkamp riffed on in his sci-fi epic District 9. A tour of the surrounding streets led by former residents is more moving than any blockbuster. *p18*

AFRICA

CAIRO

Cairo is chaos at its most magnificent, both infuriating and beautiful. Car horns, market traders and the call to prayer will serenade you as you breeze through free-to-see historic sites and museums, pausing for pocket-priced snacks and juices.

Al Azhar Mosque

Founded in CE 970 as the centrepiece of the newly created Fatimid city, Al Azhar is one of Cairo's oldest mosques, and a harmonious blend of architectural styles, spanning more than 1000 years. The tomb chamber, located through a doorway on the left just inside the entrance, has a beautiful mihrab indicating the direction of Mecca. *Gami' Al Azhar, Sharia Al Azhar; free.*

Aisha Fahmy Palace

Shuttered for years, the Aisha Fahmy Palace was built in 1907 for Egyptian aristocrat Ali Fahmy, who was King Farouk's army chief. Reopened as an arts centre, the sublime rococo interior of silk-clad and frescoed walls, carved-wood fireplaces, painted lacquerwork and wonderful stained-glass has been fabulously restored. *Mogamma Al Fonoon, Sharia Al Shaer Aziz Abaza, Zamalek; free.*

Mosque of Qaitbey

Though famously ruthless, Sultan Qaitbey was also something of an aesthete. His mosque, completed in 1474 as part of a much larger funerary complex, is widely agreed to mark the pinnacle of Islamic architecture in Cairo. The true glory is the exterior of the dome, carved with the finest, most intricate floral designs

anywhere in the Islamic world. *Northern Cemetery; free.*

Coptic Museum

There's a charge, but not a steep one, to view the 1200 or so pieces of Coptic Christian art on display in this wonderful museum, founded in 1908. There are pieces from the earliest days of Christianity in Egypt right through to early Islam. The elaborate woodcarving

Below: decorative engravings adorn the front of the Coptic Museum.

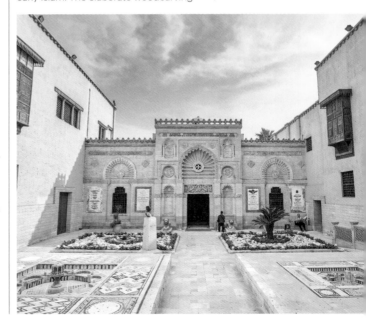

AFRICA

in the galleries is almost as impressive as the treasures they contain. *coptic-cairo. com/museum/museum.html; 3 Sharia Mar Girgis; entry LE100.*

The Tentmakers Market

The medieval 'Street of the Tentmakers' takes its name from the artisans who have produced bright fabrics used for the ceremonial tents at wakes, weddings and feasts for centuries. They also hand-make intricate appliqué wall hangings. The highest concentration of artisans is directly south after Bab Zuweila gate, in the covered tentmakers market. *Sharia Al Khayamiyya; closed Sun; free.*

Khan Al Khalili

There's no charge to duck into the skinny lanes of the Khan Al Khalili bazaar, though traders may try to separate you from your spare Egyptian pounds with extreme persuasion. Stalls arranged around small courtyards stock everything from soap powder to semiprecious stones, plus the inevitable alabaster pyramids. It's touristy in some parts, and an evocative dip into Cairo's trading past in others. *Off Sharia Al Azhar & Al Gamaliyya; free.*

Sharia Al Muizz Li Din Allah

Sharia Al Muizz was Cairo's original grand thoroughfare, and it's still powerfully atmospheric in sections. One part is devoted to the sale of *shishas* (water pipes), braziers and pear-shaped cooking pots for *fuul* (fava bean paste). Soon the stock expands to crescent-moon minaret tops, coffee ewers and other copper products. Stroll along and admire the medieval architecture mixed with Cairo's hustle and bustle. *Free.*

Ride a felucca at sunset

One of the most pleasant things to do on a warm evening is to go out on a felucca, Egypt's ancient broad-sail boat, with a supply of cold drinks and a small picnic. The Dok Dok Landing Stage is the best spot for finding a willing boatman; subject to haggling, a boat and captain should cost from LE70 per hour, plus some inevitable baksheesh. *Corniche El Nil.*

Views from Al Azhar Park

With funds from the Aga Khan Trust for Culture, a former waste dump was transformed into the city's first (and only) park of significant size. Join Cairenes on a stroll through a profusion of gardens, emerald grass and fountains, admiring the superb views over Cairo. It's most fun at weekends, when families day-trip with picnics. *azharpark.com; Sharia Salah Salem; entry LE7-10.*

Bab Zuweila

Built in the 11th century, the beautiful Bab Zuweila gate was once an execution site, and today it's the only remaining southern gate of the medieval city of Al Qahira. There are interesting exhibits about the bab's history inside the gate, while up on the roof you get panoramic vistas that stretch out to the citadel, and even better views from the top of the two minarets. *Sharia Al Muizz Li Din Allah; entry LE30,*

Mosque of Ibn Tulun

The city's oldest intact, functioning Islamic monument is ringed by high walls topped with neat crenellations that resemble a string of paper dolls. This refined structure was built between CE 876 and 879 by Ibn Tulun, who was sent to rule the region by the caliph of Baghdad. Its geometric simplicity only adds to its beauty; climb the minaret for stunning views over the courtyard and the surrounding streets of the old city. *Sharia Al Saliba; entry free*

BUDGET TRIPS TO THE PYRAMIDS

Getting to the Pyramids of Giza is easy; microbuses and bigger buses zip here from the Giza metro stop. That's where the bargains end. It costs LE120 to get into the site, then LE60 to LE300 to enter the famous pyramids, with the priciest ticket reserved for the Great Pyramid of Khufu. If you don't mind missing the interiors, the Sphinx is free to view with a site ticket, as are the tombs of nobles (rather than pharaohs) in the Eastern and Western Cemeteries.

AFRICA

MARRAKESH

A trip to Marrakesh might conjure up images of the Rolling Stones buying up half the medina, but it's easy to dip into the city's intense culture, and feast while you explore the historic streets, without dishing out all your dirhams.

✋ Mellah

Marrakesh's 15th-century Jewish quarter is a tangle of *derbs* (alleys) and mud-brick buildings. Though few Jewish families remain, their legacy survives in the Lazama Synagogue (Derb Manchoura; tip expected) and the *miaâra* or Jewish cemetery (Ave Taoulat El Miara; tip expected). Don't miss Mellah Market (Ave Houmane el-Fetouaki; free) and the artisans' showrooms at Place des Ferblantier.

✋ Musée de Marrakech

Moroccan art forms are proudly displayed within the dated but still decadent salons of the Mnebhi Palace. The central internal courtyard, a riot of cedar archways, stained-glass, intricate painted door panels and *zellige* (colourful geometric mosaic tilework), is the highlight, though don't miss the display of exquisite Fez ceramics and the palace's hammam. *museedemarrakech. ma; Place Ben Youssef; Dh70.*

✋ Bahia Palace

The entry fee has been raised in line with the painstaking restoration of this fabulous late-19th-century palace, but it's still worth it. Built over 15 years by Grand Vizier Si Moussa and his son Abu 'Bou' Ahmed, the Bahia (Palace of the Beautiful) is a 160-room complex of intricately patterned rooms, polished cedar ceilings,

cool marble and secluded courtyards, surrounded by landscaped gardens. *Rue Riad Zitoun El Jedid; Dh70.*

✋ Dar Bellarj

Flights of fancy come with the territory at Dar Bellarj, a former hospital for storks turned into Marrakesh's premier arts centre. Each year the nonprofit Dar Bellarj Foundation adopts a programme themed around living culture, ranging from film to women's textiles and storytelling. Admission is usually free (there's a charge for some events). *9-7 Toualate Zaouiate Lahdar; free.*

✋ Dar Si Said

This monument to Moroccan *mâalem*

Below: the Djemaa el-Fna square is a great place to absorb the sights and sounds of Marrakesh.

AFRICA

(master artisans) was the former home of Si Said, brother to Grand Vizier Abu 'Bou' Ahmed. Today, it's the grand setting for the Museum of Moroccan Arts, which takes visitors on a magical ride through the history and social significance of the many different forms of Moroccan carpet-making. *Derb Si Said; closed Tue; Dh70.*

Riad Kniza Musée & Galerie

Opposite the lavish *riad* of the same name, this small museum was a labour of love for its owners. When the family closed their antique shop, the overflowing collection of Moroccan heirlooms was relocated to both the *riad* and this lovely little museum, displaying High Atlas carpets, tribal jewellery and clothing, decorative Fez pottery and a wonderful collection of Amazigh sugar hammers. *riadkniza.com; 34 Derb l'Hotel; free.*

Comptoir des Mines

Once the home of a mining corporation, this building now houses Marrakesh Art Fair founder Hicham Daoudi's contemporary gallery. Restored to its Art Deco glory, the gallery hosts rotating art exhibitions profiling leading and up-and-coming African artists. Hassan Hajjaj, called the Andy Warhol of Morocco, has a dedicated gallery space. *comptoirdesminesgalerie.com; 62 Rue de Yougoslavie; closed Sun; free.*

Djemaa el-Fna

Djemaa el-Fna (Assembly of the Dead) is Morocco's largest, best-known public square, listed by Unesco for its Intangible Cultural Heritage. Created as a site for public execution in AD 1050, it is now filled by day with snake charmers, magicians and juice vendors, and at night by street-food vendors, musicians, storytellers, healers, dancers and pedlars. People-watching doesn't get any better. *Free.*

Arsat Moulay Abdeslam Cyber Park

Looking for a place to post an Instagram story? Retreat to these tranquil gardens not far from Djemaa el-Fna, where tech-head locals take advantage of the free wi-fi and the peaceful setting, combining lemon, olive and pomegranate trees, water features and neat paths. *Ave Mohammed V; free.*

Koutoubia Mosque & Gardens

Koutoubia, built in the 12th century on the site of an earlier mosque imperfectly aligned with Mecca, is one of Marrakesh's great sights. The interior is off-limits to non-Muslims, but everyone can admire the 70m-high minaret, topped with gleaming copper balls – best viewed while strolling through the peaceful, palm-strewn gardens that surround it. *Cnr Rue el- Koutoubia & Ave Mohammed V; free.*

The Tanneries

If you can stand the horrendous smell, then the outdoor tanneries are one of Marrakesh's classic free sights. Animal hides are treated in hundreds of concrete vats, following medieval techniques that involve quicklime, water, blood and even pigeon poo. Producing the many-hued leathers sold in the city's souqs is dirty work and often uses dubious chemicals, but hundreds still take on the job. *Ave Bab el-Debbagh; free.*

Le Jardin Secret

This historic *riad* was once owned by powerful *qaid* (local chief) U-Bihi, but it's the traditional Islamic garden that is so special, fed by a restored original *khettara* (underground irrigation system). The bare chambers include excellent exhibits on the building's history and the role of gardens in Marrakesh culture. *lejardinsecretmarrakech.com; 121 Rue Mouassine; Dh60.*

MEDINA ETIQUETTE

Marrakesh has always been a mercantile metropolis, but come prepared if you want to maximise your dirham. Hard sell is commonplace in the souqs, so be ready to stand your ground. Stallholders are willing to haggle; just don't get involved in protracted negotiations if you don't plan buying. You might also get the idea that guides are trying to edge you towards certain stalls – of course they are – so feel free to politely insist on looking elsewhere.

CAPE TOWN

South Africa's proudly multicultural Mother City sprawls around Table Mountain and down towards the Cape of Good Hope. Tear yourself away from mountain and beaches to explore fascinating neighbourhoods and low-cost museums that tell the city's story.

✋ District Six

Once a tight-knit, multicultural community, District Six was transformed into a whites-only zone under apartheid. By 1966, 60,000 occupants had been evicted. Don't miss the District Six Museum (districtsix. co.za; 25a Buitenkant St; closed Sun; R45), which also runs Sunset Walking Tours (R110) led by ex-residents.

✋ St George's Cathedral

Even if a guided trip to Robben Island (robben-island.org.za; R600) is beyond your budget, don't miss the 'People's Cathedral', where Archbishop Desmond Tutu led a dignified decades-long struggle against apartheid. The main service is at 9.30am on Sundays. *sgcathedral.co.za; 5 Wale St; free.*

✋ Whatiftheworld Gallery

Occupying a space in a decommissioned Woodstock synagogue, this gallery exhibits contemporary Southern African artists' work. *whatiftheworld.com; 16 Buiten St, Woodstock; closed Sun & Mon; free.*

✋ Guga S'Thebe Arts & Cultural Centre

Decorated with polychromatic murals, this is an impressive township building with its own theatre. Come to see pottery workshops or performances by local groups telling township stories. *capetown.gov.za; cnr King Langalibalele Ave & Church St; free.*

✋ Iziko Museums

This 11-strong collection of museums and attractions includes the South African National Gallery, South African Museum and Planetarium and Maritime Centre. Entry is normally about R30, but for eight or nine days each year, typically public holidays, they all throw their doors open for free; see the website for details. *iziko.org.za*

✋ Walk Bo-Kaap

Bo-Kaap, the former Malay Quarter, is famous for its multicoloured streets, mosques and food. Wander freely around the streets, visit the inexpensive Bo-Kaap Museum (iziko.org. za; 71 Wale St; closed Sun; R20), then grab a drink nearby on Long St, focus for the city's buzzy entertainment scene.

Below: Chiappini Street in Bo-Kaap; it's thought that while the houses were leased to enslaved people, the buildings had to be white. When those restrictions were lifted, the owners painted their homes as an expression of freedom.

© Ariadne Van Zandbergen | Lonely Planet

AFRICA

TUNIS

The Tunisian capital is more than just its magnificent medina: waves of colonisation have endowed the city's fabric and culture with a rich and complex flavour, best experienced for free (or penny prices) in its hammams and mosque-lined backlanes.

✋ Bardo Museum

The moderate entry fee at Tunisia's top museum is a small price to pay for its magnificent collection of Roman mosaics. The massive collection is housed in an imposing palace complex built under the Hafsids (1228–1574), and fortified and extended by the Ottomans in the 18th century. *www.bardomuseum.tn; closed Mon; entry 11DT.*

✋ Wander the Medina

A maze of vaulted souqs, alleys lined with massive, brightly painted doors opening to lavishly tiled *dars* (mansions), and landmark mosques with distinctive octagonal minarets, the Tunis Medina is deservedly a Unesco World Heritage Site. Some monuments charge for entry, but walking the streets is gratis – and highly evocative. *Free.*

✋ Hammam El Kachachine

One of the medina's oldest still-functioning bathhouses, this historic place offers a fully authentic hammam experience. You'll need to take a towel (or *fouta*, a cotton bath sheet), and soap. It's men only, and entry is through the old-fashioned barber store. *30 Rue des Libraires; self-service wash 3DT, massage 5DT.*

✋ La Malga Cisterns

The restored remnants of the huge 2nd-century-CE cisterns that housed Roman Carthage's water supply are located at the foot of Byrsa Hill. The original complex was nearly 1km long, fed by a huge aqueduct carrying mountain spring water from Zaghouan. *Rte de Carthage; free.*

✋ Galerie El Marsa

Occupying a beautiful vaulted space in the heart of the La Marsa neighbourhood, this is one of Tunisia's most respected commercial galleries, with a small but impressive stable of contemporary artists from the Arab world. Swing by and see who is currently stirring up a storm. *galerielmarsa.com; 2 Pl de Safsaf; closed Sun; free.*

✋ Dar Lasram

Former home of the Lasram family, who provided Tunisia's *beys* (rulers) with scribes, this was one of the first historic mansions restored by Association de Sauvegarde de la Médina de Tunis. The interior features magnificent, richly tiled rooms and courtyards, and displays documenting the Association's work. *asmtunis.com; 24 Rue de Tribunal; closed Sun; free.*

FOOD & DRINK

Africa isn't famous for food-related travel, but there are feasts here to rival the world's top gourmet hubs if you know where to look. Finding good food on a budget is more of a challenge: locals subsist primarily on meat or fish with rice or 'starch' (any of a dozen types of steamed or mashed grains and roots) and sauces. You may feel compelled to splash out on the odd expensive sit-down meal just for variety.

STREET FOOD

Street food in Africa is a bit more rough-

Above: grapes growing in Stellenbosch, soon to become great-value wine. Right: try local food, such as this baked tilapia.

and-eady than you may be used to. The favourite street treat is grilled meat of indeterminate origin, roughly chopped and cooked over coals on a half oil drum.

When it comes to regional delicacies, there's budget gold to be found on the streets: *bobotie* (minced beef topped with egg) in South Africa; *kelewele* (spiced fried plantain) in Ghana; *suya* (peanut-crusted grilled meat) in Nigeria. Ask someone local to point you to the best stalls and specialties.

• **Bake off** – thrift-friendly bakeries do a brisk trade, with a tangible French influence in Morocco, Tunisia and Algeria, Portuguese hints in Mozambique, and a Dutch twang in South Africa.

• **Cheap market eats** – food markets (particularly the old town souqs of North Africa) invariably have food stalls serving up cheap feasts.

• **Farmers' markets** – deli-style farmers' markets are big news in South Africa and Namibia, with all the expected global food stands.

SIT-DOWN DINING

If you get used to local fare and local restaurants, you can eat out very cheaply in Africa, particularly if you develop a taste for *ugali* (maize dough) and *fufu* (plantain or cassava dough). Most towns and traveller hubs have a scattering of upmarket cafes and restaurants, serving familiar global tastes – pizzas, burgers and fries, grilled meats, sandwiches – for a notable shift up in price.

• **Avoid supermarkets** – prices can be panic-inducing; shop instead at local produce markets to keep a lid on self-catering costs.

• **Eat local** – conquer any nerves about unfamiliar foodstuffs and you'll find feasts for pennies; start the search near the local church or mosque, catering to post service congregations.

• **Dine near water** – find cheap seafood at fish-landing beaches, harboursides and coastal markets. Inland, along lakeshores and riverbanks, grilled tilapia fish and rice is a simple feast you'll become very attached to.

• **Big lunch, small dinner** – outside North Africa and southern Africa, lunch is often a bigger meal than dinner; after dark, the best option is often bar snacks with a beer.

SMART DRINKING

A cold beer at the end of a dusty bus ride won't tip your finances into the red, and even fine wine doesn't need a platinum card if you drink close to source in South Africa.

• **Drink like a local** – many African countries skip alcohol for religious reasons; substitute with locally-grown tea and coffee and fabulous juices.

• **Be ambitious** – Africa has its own unique (and inexpensive) beverages; try papaya wine, *mbege* (fermented bananas), *boukha* (Tunisian fig brandy) or *umqombothi* (maize beer).

HIGHLIGHTS

BOSCHENDAL, FRAN-SCHHOEK, SOUTH AFRICA
The full South African wine experience: an historic estate sprawling over stunning countryside, and a mix of elegant and informal dining options, meaning backpackers can sip Pinotage and snack on deli sandwiches while the high fliers go for all-out feasts. *p22*

BISSARA IN FEZ MEDINA, MOROCCO
When roaming Fez's ancient medina, set yourself up for the day with a hearty bowl of *bissara* (fava-bean soup), served with a hunk of bread and a splash of olive oil, and livened up with cumin, chili and salt. Oh, and did we mention it costs pennies? *p25*

AFRICA

CAPE WINELANDS

Wine isn't a fad in South Africa's cape – they've been fermenting fine vintages since 1659, and many Cape vineyards offer inexpensive tastings, tours and more. Good drinking is backed up by good eating, and often for less than you might expect in wine country.

⚑ Boschendal, Franschhoek

This is a quintessential Winelands estate, with gorgeous architecture, food and wine. As well as tastings, there are excellent vineyard and cellar tours; soak up the grapes with sandwiches at the Farmshop & Deli, or join the big Sunday buffet lunch (R325) in the 1795 homestead. *boschendal.com; Groot Drakenstein, Rte 310; tastings from R45, tours R50-80.*

⚑ Leopard's Leap, Franschhoek

Leopard's Leap's bright, modern, barn-like tasting room has comfy couches strewn around, or you can sit at the bar for a slightly more formal quaffing. The rotisserie restaurant (Thu-Sun) serves meaty feasts by weight; budget for around R150 per person. *leopardsleap. co.za; Rte 45; closed Mon; tastings from R55.*

⚑ Fyndraai Restaurant, Franschhoek

This interesting restaurant serves original dishes inspired by the Cape's varied cultures, prepared using herbs from the on-site indigenous garden. Expect some surprises. A handy glossary on the menu helps you to learn some traditional foodstuffs. *solms-delta.com; Delta Rd, off Rte 45; mains R140-185.*

⚑ De Villiers Chocolate Cafe, Franschhoek

The excellent single-origin chocolate here is made at a boutique chocolatier in Paarl and used in all the cakes, pastries and ice cream on offer. There's also excellent coffee; try the ice cream or coffee tasting (R55). *devillierschocolate.com; Heritage Sq, 9 Huguenot Rd; pastries from R30.*

⚑ Babylonstoren, Paarl

After a tasting at this sprawling winery and fruit farm, explore the formally designed garden. Inspired by Cape Town's Company's Garden, it is an incredible undertaking, featuring edible and medicinal plants, lotus ponds, espaliered quince trees and a maze of prickly-pear cacti. *babylonstoren.com;*

Below: there's an entrance fee to visit Babylonstoren farm's amazing gardens, which date from the 17th century.

AFRICA

Simondium Rd, Klapmuts; entry R10, tastings R30.

Blacksmith's Kitchen, Paarl

The inexpensive eatery at the Pearl Mountain Winery is a simple, hearty treat. Grab a table under the trees and enjoy an unfussy lunch of wood-fired pizza or roast pork belly with views of the vineyards, town and mountains beyond. pearlmountain.co.za; Pearl Mountain Winery, Bo Lang St; closed Mon; mains R65-195.

Tea Under The Trees, Paarl

The only downside to this fabulous tea garden is that it's only open for half the year. Based at an organic fruit farm, it's a wonderful place to sit under century-old oak trees and enjoy an al fresco cup of tea, a light lunch or a large slice of home-baked cake. http://teaunderthetrees.co.za; Main St, Northern Paarl; Oct-Apr, closed Sun; mains R40-55.

Excelsior, Robertson

At Excelsior, free tastings take place on a delightful wooden deck overlooking a reservoir. The real draw, though, is the 'blend your own' experience, where you can mix wine varieties and take home a bottle of your own creation. The restaurant serves great roosterbrood (traditional bread cooked over coals) sandwiches. excelsior.co.za; off Rte 317; closed Sun; tastings free.

Schoon Croissant & Coffee Bar, Stellenbosch

Grab coffee and a pastel de nata to go at this hatch in the town centre, or sit in the courtyard and tuck into a filled cronut or a decadent croissant topped with chocolate and nuts. For a wider selection head to the Schoon Manufactory (91 Bird

St, La Colline; closed Mon) north of town. schoon.co.za; De Wet Centre, Church St; closed Sun; croissant R45-75.

Viljoensdrift, Robertson

One of Robertson's most popular places to sip, especially on Saturday. Put together a picnic from the deli, buy a bottle from the cellar door and taste on an hour-long boat trip along the Breede River (R90). Boats leave on the hour from noon; bookings essential. viljoensdrift.co.za; Rte 317; closed Sun; tastings free.

Villieria, Stellenbosch

Villiera produces several excellent Méthode Cap Classique sparkling wines and a highly rated and well-priced shiraz. Excellent two-hour wildlife drives (R260) visit antelope, zebras, giraffes and various bird species on the estate, and are followed by a tasting. www.villiera.com; Rte 304 & Rte 101; closed Sun; tastings R40.

Vergelegen, Stellenbosch

The van der Stel family first planted vines here in 1700 on an estate with ravishing mountain views and a 'stately home' feel. You can take a tour of the gardens and cellar or just enjoy a tasting of four of the estate's wines (the flagship Vergelegen Red costs an extra R10). vergelegen.co.za; Lourensford Rd, Somerset West; tastings from R30, tours R50.

Stone Kitchen, Wellington

On a gravel road 5km outside Wellington, Stone Kitchen is part of Dunstone Country Estate, an unpretentious, family-run winery that makes for a fun, family friendly day out. The menu features filling classics like quiche, burgers, pies and ribs, and you can enjoy a wine tasting at your table while you dine. dunstone.co.za; Bovlei Rd; mains R75-145.

AFRICA

JOHANNESBURG

Known as Jo'burg or Jozi to its buds, this rapidly changing city is the vibrant heart of South Africa. It's famously friendly, fun and foodie, with a multicultural menu, inexpensively fusing African, Asian and European themes.

✋ Breezeblock

The designers have done a grand job transforming this old Chinese restaurant into a pastel-shaded haven of mid-century cool in gritty Brixton. Many come for delicious breakfast dishes, from huevos rancheros to baked pancakes, and find themselves lingering till lunch over coffee. *facebook.com/ pg/breezeblockbrixton; 29 Chiswick St, Brixton; closed Mon; mains R26-65.*

✋ Bergbron Plaaskombuis

Modelled after a traditional *padstal*, or roadside stall, Bergbron Plaaskombuis serves authentic Afrikaans cooking at rock-bottom prices. Try the breakfast *skilpadjies* – balls of minced liver with a side of fried eggs. *facebook. com/Bergbron.Plaaskombuis; 268 Weltevreden Rd; closed Mon; mains R25-70.*

✋ Patisserie de Paris

Owner Paul Zwick spent 30 years in the television business before training as a chef in France. His patisserie produces delectable French breads and pastries, served with posh cheese and pâté from France. Eat in or grab a bag to take away. *patisseriedeparis.co.za; 9 Mackay Ave, Blairgowrie; mains R50-100.*

✋ Urbanologi

Even if Mad Giant's very quaffable beers are not your thing, don't pass by the microbrewery's fabulous restaurant. You really can't go wrong with yakitori chicken with spicy chimichurri sauce or teriyaki poached carrots with kumquat marmalade. *madgiant.co.za; 1 Fox St, Ferreiras Dorp; mains R60-80.*

✋ Cheese Gourmet

It's dairy all the way at the Cheese Gourmet. The shop part sells a dizzying selection of South African cheeses, olives, breads and spreads, while the adjoining cafe serves a fantastic cheese-themed menu. Staff offer solid advice and generous tastes of each cheese for try-before-you-buy types. *71 7th St, Linden; closed Sun; mains R50-100.*

✋ Momo Baohaus

Minimalist decor greets visitors to Momo Baohaus, where fusion Asian tapas is the order of the day. Dine on crispy pork dumplings, sushi rolls and the fluffiest bao buns followed by veggie bowls that are as colourful as they are flavourful. There's no liquor licence, but you can bring your own. *momobaohaus.com; 139 Greenway, Greenside; closed Mon; mains R52-90.*

AFRICA

FEZ

Historic, cultured Fez is like stepping into a North African folk tale, with medieval magic crackling in medina lanes passable only on foot (human or donkey). Add in fabulous, thrifty feasting, and it's easy to see the appeal.

Bissara stalls in the medina

Try the Fassi speciality of *bissara* (fava-bean soup), served from tiny shops in the medina. Perfect fuel for exploring the city, the soup is served with bread and a dash of olive oil; season with salt, cumin and chilli. *Search around Talaa Kebira. Soup Dh6.*

Chez Abdellah

This is a tea shop-cum-theatre in which the eponymous Abdellah, a born showman, mixes teas to order – not just mint, but also wormwood and other traditional digestives. The secret is the water: it comes from the holy Kairaouine complex. *Rue Lmachatine; closed Fri; tea prices vary.*

Café Clock

Crossed by a maze of stairs, Clock is a prime spot to rest and nourish yourself. The menu flips between Moroccan and European: a signature camel burger, ras el hanout-spiced potato wedges and interesting vegetarian options. Free cultural events include storytelling, live music, movies and more. *cafeclock.com; Derb El Magana, Talaa Kebira; mains Dh60-85.*

Famille Restaurant Berrada

'Famille restaurant' says it all. Everything is traditional and local, and diners are invited into the kitchen to taste the day's selections before ordering (there's no written menu).

All meals come with bread and salad. *57 Sagha El Achebine; closed Fri, mains Dh50.*

Ruined Garden

Dine in the garden in summer or by a fire in winter at this aptly-named spot. Lunch includes stews and 'tapas' (actually Moroccan salads), while dinner has a few more elaborate dishes. *ruinedgarden.com; 13 Derb Idrissi; tapas Dh20, mains Dh70-100.*

Le Tarbouche

This hip snack joint has a creative menu. The Moroccan tabbouleh goes down well with a rosemary-mint lemonade. For mains, there's spicy *merguez* sausage pizza and a veggie *bastilla* (filo pie). *facebook.com/letarbouchefes; 43 Talaa Kebira; snacks Dh25-30, mains Dh45-65.*

Below: exploring the medina and browsing the shops, such as these at Bab Boujloud Gate, is a great (free) way to get to know Fez.

AFRICA

BEST
AFFORDABLE
WILDLIFE ENCOUNTERS

Seeing Africa's wildlife Big Five (and many more) is the stuff of travel dreams. Here are some places to make that dream affordable.

NAIROBI NATIONAL PARK, KENYA

Considering that the biggest cost for a safari is transport, you could do worse than take a hire car to this teeming reserve, which unfolds just 10km from central Nairobi. Giraffes are a definite spot and you might also see lions, leopards, cheetahs and rhinos. *kws.go.ke/parks/nairobi-national-park; entry US$43.*

GORILLAS IN THE DEMOCRATIC REPUBLIC OF THE CONGO

Getting close to gorillas is never cheap, but for a once-in-a-lifetime experience, it's right up there. Skip the expensive permits in Rwanda and Uganda and head to the DRC, where gorilla permits start from US$400 for treks into Kahuzi-Biéga national park. *kahuzibiega. wordpress.com.*

WHALES FOR FREE, SOUTH AFRICA

The coast of South Africa is thronged by whales from June to December; boat trips will get you close for a price, but there's no charge to watch from land. The coastal path at Hermanus offers regular sightings, or consider the small investment to access the cliffs at Robberg Nature & Maritime Reserve at Plettenberg Bay. *capenature. co.za; entry R40.*

KRUGER SELF-DRIVE, SOUTH AFRICA

When it comes to cheap self-drive safaris, we rate Kruger for reasonable costs and ease of access. Agencies at Kruger Mpumalanga International Airport, Hoedspruit (Eastgate) Airport and Kruger Park Gateway Airport can rent you a car (no 4WD required), then all you need to pay to see the Big Five is the entry fee. *sanparks.org/parks/ kruger; entry R424.*

BABOONS FROM THE BUS, KENYA

The road from Nairobi to Mombasa and the coast slices right through the middle of Tsavo National Park, and public buses have to slow down regularly as troops of baboons mob passing vehicles in the hope of handouts from passengers. It's a free-for-all, so keep your fingers inside the vehicle. *Nairobi-Mombasa bus KSh700.*

BUDGET SAFARIS, GHANA

Known for its elephants, Mole National Park also contains buffaloes, kob antelopes, baboons, warthogs and numerous bird species. Share the park's hire vehicles (C80 per hour) with others to shrink costs, or join organised morning and afternoon safaris for C20-60. There are budget rooms and camping too. *molenationalpark. org; Mole NP, Ghana; entry C40.*

SHARKS WITHOUT THE BITE, SOUTH AFRICA

Cage diving with great white sharks is a thrill that comes with a financial bite – expect to pay R1700 upwards for a dive with a Fairtrade-accredited operator from Gansbaai in South Africa. A less costly option is to join a seal-spotting trip with Simon's Town Boat Company to Seal Island, where great whites can often be seen leaping from the water. *boatcompany. co.za; trips from R450.*

DESERT SAFARIS AT ETOSHA, NAMIBIA

The desert-like environment around the massive Etosha salt pan is captivating, and it forces wildlife out into the open at the park waterholes. Great roads mean small rental cars are an option, and cheap camping adds to the park's budget credentials. *etoshanationalpark. org; entry N$80, car fee N$10.*

FESTIVALS & EVENTS

At first glance, many of Africa's festivals appear to be free to all, particularly cultural and religious fairs, but once you factor in transport and accommodation in remote locations, a tribal fair can cost as much as a rockstar weekender. Camping and travelling independently are the best ways to save money.

PRACTICALITIES

Africa's top festivals feature spectacularly photogenic costumes, dance, music and rituals – something heavily marketed by tour agencies. Plan travel well ahead,

Above: Lalibela Church in Ethiopia. Right: taking in a local sports event will be less expensive than catching the Springboks.

ideally with your own vehicle so you aren't held hostage by logistics.

- **Camp out** – many events in Africa are in tiny, remote outposts that see half a million people at festival time, and nobody else all year; often, camping is the only option.
- **Travel under your own steam** – don't rely on being able to get to festivals by public transport; buses, trains and trucks are stuffed like sardine cans, and you'll have a much easier time with your own wheels.
- **Free culture in the city** – unlike in many regions, it's the big cities that are most likely to lay on free performances; Marrakesh, Fez, Tunis, Casablanca and Cape Town are good bets.

RELIGIOUS FESTIVALS

Religion is at the heart of the festival calendar in Africa. Islam, Christianity, Hinduism (a legacy of colonial times) and local religions, from Ethiopia's Waaqeffanna to Benin's voodoo, all stand up and demand to be heard at festival time.

During major faith events such as Ramadan, Christmas and Easter, people relocate in vast numbers and many businesses close, meaning logistical problems reaching some festivities by public transport; again, a hire vehicle will cut costs and hassle.

MUSIC & CULTURE

Africa's music festivals cover a lot of ground. You'll find the usual international-style rock, pop and jazz extravaganzas (with stratospheric ticket prices) but also bargain-priced celebrations of local music and dance, such as the Amini Festival in Democratic Republic of the Congo.

- **Small is beautiful** – Africa's biggest tribal festivals attract heavy tourist crowds; smaller, regional festivals have less of a scrum for transport, accommodation and vantage points for taking photos.
- **Cultural encounters** – many cultural and tribal festivals are in remote, inaccessible locations, but going as part of a tour can be expensive; instead, drive yourself and camp (sites are often included in ticket prices).
- **Early bird specials** – many African cultural festivals charge higher fees for foreign visitors, but discounts for early booking can make up some of the difference.

MUSIC & CULTURE

There's plenty of gourmet action in Southern Africa and North Africa, but few foodie extravaganzas in between, though Lagos, Nairobi and Accra are bucking the trend. Most food fests are focused on restaurants and vineyards; there are bargains to be had at Cape Town's 10-day Street Food Festival (streetfoodland.co.za; Sep) and the Stellenbosch Wine Festival (stellenboschwinefestival.co.za; Jan/Feb).

SPORTS

Africa's sporting calendar is followed passionately by locals; with so many countries competing for league trophies, you might actually be able to get a ticket for some international football games at a reasonable price. We can't say the same for South African rugby or cricket.

HIGHLIGHTS

AMANI FESTIVAL, GOMA, DRC
A pocket-pleasing US$1 per day is all it takes to see the best of music, dance, comedy and more from Democratic Republic of the Congo, Rwanda and the surrounding region for this three-day celebration of peace between neighbours, held every February in Goma. *p30*

OUIDAH VOODOO FESTIVAL, BENIN
In Benin, the Fon people practise the original unadulter-ated version of voo-doo, and 10 January is the biggest date in the calendar year. Devotees wear vivid costumes, consult the spirits and haggle over fetishes (devotional objects) in festival markets, alongside dancing, drumming and drink-ing. *p31*

AFRICA

FESTIVALS

🖐 Amani Festival, Goma, DRC

You don't get much these days for US$1, except perhaps a day of top Congolese music and dance. Conceived to promote peace between the DRC and its neighbours after the violence of the 1990s, the festival offers three days of vibrant regional culture, from pop and rap acts to stand-up comedy and traditional dance. *amanifestival.com; Feb; per day US$1.*

🖐 Cape Town Carnival, South Africa

Born on Long St amid the orgy of euphoria and energy that surrounded SA's hosting of the 2010 Football World Cup, this carnival has since grown into a massive annual gathering to celebrate South Africa's diversity. Some 50,000 people watch and participate in festivities that engulf Green Point, with floats, music, dancing and after-partying. *capetowncarnival.com; Fanwalk, Green Point; Mar; free.*

🖐 Epiphany, Ethiopia

Some 63% of Ethiopians are Christian, and Biblical feast days are a huge deal across the country. Known locally as Timkat, Epiphany is probably the biggest celebration, marking the baptism of Jesus in the River Jordan, and huge numbers of people dress in white for ritual processions and recreate the event by leaping into the blessed waters of Fasiladas' Bath at Gonder. *19 Jan; free.*

🖐 Festival of Popular Arts, Marrakesh, Morocco

This midsummer festival lights up the streets of Marrakesh every July.

International performers join Berber minstrels, Gnawa musicians, sword-swallowers, acrobats, dancers, snake charmers and street performers from across Morocco. Performances can be seen at outdoor venues across the city over the festival's 10 days. *morocco.com/theater/marrakech-popular-arts-festival; July; free.*

🖐 Hermanus Whale Festival, South Africa

The return of southern right whales to the coast of South Africa is when Hermanus town lets its hair down. Whale-watching trips are the focus, but there are also film screenings, music, comedy and street performances to seal the deal. Much is free, including whale spotting from the Hermanus cliff path. *hermanuswhalefestival.co.za; Sep; prices vary.*

🖐 Knysna Oyster Festival, South Africa

Oysters by the plate-load are just part of this lively celebration on South Africa's

Below: a southern right whale and her calf come close to the shore at Hermanus, South Africa.

garden route. There are live shows, cooking and shucking competitions, cycle races, a forest marathon, talks with sporting celebs and more. Much is free, though some events are ticketed. It's one of the Cape's biggest events. *knysnaoysterfestival.co.za; Jun/Jul; event prices vary.*

⚝ Lake of Stars Festival, Malawi

Lake Malawi is the venue for this bumper serving of African music, theatre and culture. There's something for everyone here: poetry, fashion, art, even yoga sessions on the banks of the lake. Ticket prices are fairly high, but include camping (bring your own gear) in this wonderfully remote location. *lakeofstars.org; Sep/Oct; three-day tickets from US$75.*

⚝ Maha Shivaratri, Mauritius

Mauritian Hindus believe that Ganga Talao, the crater lake at Grand Bassin, was formed when Shiva let drops of the River Ganges fall to earth. Accordingly, for the festival marking Shiva's cosmic dance, Hindus walk barefoot from across the island carrying colourful bamboo floats and pay homage at the sacred lake. *Feb/Mar; free.*

⚝ Marsabit Lake Turkana Cultural Festival, Kenya

More than a dozen tribal communities from Lake Turkana come together for the Marsabit Lake Turkana Cultural Festival, a three-day celebration of tribal music, dance and culture. The biggest cost is travel and accommodation; come by self-drive and camp to keep overheads down. *.en-gb. facebook.com/ltcfmarsabit; Jun; costs vary.*

⚝ Mawazine, Morocco

Held in Rabat, Morocco's biggest music festival attracts legions of international stars, of the calibre of Nile Rodgers, Stevie

Wonder and Ellie Goulding, alongside lots of local talent from across Africa and the Middle East. But the best part of all is that many shows can be seen for free, with only the areas closest to the stage reserved for ticket holders. *mawazine.ma; Jun; free, tickets for some areas.*

⚝ Ouidah Voodoo Festival, Benin

This annual celebration is a nationwide holiday in Benin, but it's at Ouidah that the vodun (voodoo) vibes are released with the most passion. Presided over by a *feticheur* (high priest) from the Fon community, the festival attracts delegations of villagers in surreal costumes, depicting clan spirits, with frenetic markets selling religious fetishes and animal sacrifices. *10 Jan; free, but most visit on organised tours.*

Sauti za Busara Swahili Music Festival, Zanzibar, Tanzania

Atmospheric Stone Town in Zanzibar is the appropriate setting for this celebration of Swahili culture and music. Spread over two days, the music is likely to blow you away, so it's well worth the ticket price (just try not to think about the fact that locals pay just US$4). *busaramusic.org; Feb; tickets for international visitors US$60.*

Above: making offerings to the Gods during Maha Shivaratri Hindu festival in Mauritius.

AFRICA

© Anna Schlosser | Shutterstock

OUTDOORS & ADVENTURE

Most people head to Africa with one thing in mind: getting out into the wild and close to the wildlife. Going to Africa without taking at least one safari would be like going to Vegas without stepping into a casino – unthinkable. Nature provides other thrills too, often for free:

rock climbing, snorkelling, trekking and more, so climb into your khakis and head out to explore.

NATIONAL PARKS & NATURE RESERVES

Africa's grasslands and forests throng

Above: bring or borrow a board to surf in South Africa, here at Lubanzi. Right: there are ways to spend less on the safari of a lifetime.

with elephants, zebras, giraffes, apes and monkeys and more kinds of antelope than you'll ever be able to remember the names of, (literally) followed by hungry lions, leopards, cheetahs and other top-of-the-food-chain predators.

The bad news? Getting into most of Africa's wilderness reserves involves a hefty outlay. As well as national park entry fees of between US$20 and US$100 per day, you'll need a jeep and guide (or a rental 4WD) and accommodation in wilderness areas. Even split between a group, the costs add up.

To save costs, try the following tips:
• **Pick your park** – if you don't mind alternative wildlife to the Big Five, seek out less well-known parks with smaller entry fees.
• **Be a happy camper** – tented camps cost less than bricks-and-mortar lodges, and they have more atmosphere. To save even more, bring your own camping equipment to wildlife-safe campgrounds.
• **Stay outside** – staying inside a reserve

means wildlife on demand, but a big mark-up; cheaper accommodation can often be found just outside the gates.
• **Share the load** – safaris assembled on the ground (rather than booked from outside the country) are priced per vehicle; get a group together and costs-per-person (apart from park entry fees) shrink.
• **Self-drive** – even with an extra vehicle fee on top of park entry (except in South Africa), hiring your own vehicle can still be the cheapest way to safari.
• **Get a pass** – South African National Parks' Wild Card (sanparks.org/wild_new) gives unlimited access to 80 national parks across Southern Africa for a year from R720.

ADVENTURE ACTIVITIES

Africa's adventures are concentrated on but not limited to the national parks. You'll find plenty of top locations for surfing, scuba diving, rock climbing, trekking and more, particularly in the far north and far south. With the wildlife and tough terrain, many are willing to pay extra for an organised trip; it's possible to do things independently with your own gear, but consider a local guide for safety.
• **Surf safely** – bring your own board for free breaks across Africa; follow local shark advice – they're out there!
• **Snorkel for free** – bring your own gear to the Red Sea coast of Egypt for limitless free snorkelling right off the beach.
• **Independent peaks** – Kilimanjaro isn't open to independent trekkers; to go it alone, try Mount Kenya, Morocco's Atlas Mountains or South Africa's Drakensberg.
• **Dramatic drives** – a rented 4WD (or, for the brave, a bicycle) opens up thousands of miles of free adventures: consider Namibia's Skeleton Coast; South Africa's Chapman's Peak; or the Atlas Mountains.

HIGHLIGHTS

DRIVE THE TIZI N'TEST PASS, MOROCCO

The definitive Atlas Mountains drive, the pass joining Marrakesh and Taroudant strains over the 2100m Tizi n'Test pass, zig-zagging through epic Atlas scenery. You can do the trip in a conventional car, meaning no expensive 4WD hire costs. p34

LOWER ZAMBEZI NATIONAL PARK, ZAMBIA

See most of the Big Five for less on a self-drive safari through Lower Zambezi National Park, where the daily entry fee is US$25, compared to over US$70 for the top national parks. Along the river banks you'll spot plenty of big predators at a bargain price. p37

AFRICA

ATLAS MOUNTAINS

With landscapes as dramatic as those served up in the Atlas Mountains, you won't mind investing the dirhams for a guide to take you to the best spots. But much is free to see and do in North Africa's favourite mountain playground.

🖐 Mountain pass driving on the Tizi n'Test

Blasted through the mountains by the French in the late 1920s, the awe-inspiring road over the Tizi n'Test pass (2100m) between Marrakesh and Taroudant was the first modern route linking Marrakesh with the Souss plain. With a rental car, its hair-raising hairpin bends offer one of the most exhilarating drives in the country. *Free.*

🖐 Ancient petroglyphs at Oukaimeden

For a free peek at the Atlas' ancient history, a sign near the entrance to Oukaimeden marks the way to a rocky outcrop carved with illustrations of pastoral life and geometric patterns from 4000 years ago. An unofficial guide will inevitably wander over to show you around for a small tip. *Free (tips welcome).*

🖐 Trails and waterfalls in Imlil

The villages around Imlil are framed by incredible landscapes, with hiking trails radiating on all sides. This is the main trailhead for Jebel Toubkal, but escape the crowds by trekking southwest over Tizi n'Mzik (2489m) to the wonderful Cascades d'Irhoulidene waterfall. The Bureau des Guides arranges guides at fair prices. *bureaudesguidesimlil.com; Dh300/400 per half/full day.*

🖐 Hike or ride from Todra to Boumalne Dades

Walks around Todra range from easy, free rambles through palm groves to multiday treks over rough piste (dirt track) all the way to the Dades Gorge. Most hotels in the gorge and Boumalne Dades can connect you with hiking guides, or you can book horseback trips at Auberge Cavaliers. *facebook.com/equestriantrep; horse-riding per hr €15, hiking guides per day Dh300.*

🖐 Rock-climbing Todra Gorge

Todra's vertical rock faces offer sublime rock-climbing routes (French grades 5 to 8). Pilier du Couchant, near the entrance

Below: mountain biking is a well established activity in the Atlas Mountains; either join a tour or, with a bit more planning, bring a bike and go independent.

© Francois Seuret | Shutterstock

to the gorge, offers classic long climbs, while the Petite Gorge is better for novice climbers, with good short routes. It's free with your own gear, or reliable equipment can be hired from Aventures Verticales. *climbing-in-morocco.com; gear hire half/ full day Dh125/200.*

Climb Jebel Toubkal

North Africa's tallest peak, Jebel Toubkal (4167m) doesn't require technical climbing experience. In summer, anyone in good physical condition can reach the summit. Allow two to three days to conquer the mountain, making allowances for changeable weather. A guide is needed to continue above Aroumd; the Bureau des Guides in Imlil can arrange guides for Dh300/400 per half/full day. *bureaudesguidesimlil.com.*

Get off the trail at Zaouiat Ahansal

Fantastically remote Zaouiat Ahansal was founded in the 13th century by travelling Islamic scholar Sidi Said Ahansal. With the tarmac road from Azilal arriving in 2013, the village has been making a name for itself among serious climbers and adventurous trekkers who don't want to share the experience with a crowd; the sheer cliff of Tagoujimt n'Tsouiannt is one challenging ascent. *Free.*

Reserve Naturelle de Tamga

This vast national reserve covers over 84 sq km of rolling forested mountains. Birdwatchers will have a field day (or several) observing over 100 species, while climbers will be left stunned by their first sight of the Cathédrale des Rochers massif. However, this is territory for self-starters; there are no marked trails, no on-site signage or interpretation, and nobody to tell you where to go. *Free.*

Take a selfie at Cascades d'Ouzoud

The many-tiered Cascades d'Ouzoud waterfalls are stunningly beautiful, with several distinct steps, the largest featuring a massive 100m drop. It's one of the most popular day trips from Marrakesh so you won't have this natural idyll to yourself, but it's still a stunning spot and well worth a visit. *Free.*

Walk under Imi n'Ifri

This natural travertine bridge formed over a gorge 1.8 million years ago; according to legend, it represents two local lovers whose families kept them apart, so they held hands and turned to stone. You can walk down into the gorge and through this toothy maw unaccompanied – the paths are clearly marked by the bridge. *Free.*

Ski Oukaimeden

From December to April visitors come to Oukaimeden to ski on seven runs (nursery to black) covering a total of 10km, accessed by seven tows and by the highest ski lift in Africa (3268m). Gear, passes and lessons are available in town at prices that will delight those used to European and North American rates. Cross-country skiing is also available. *oukaimeden.org; prices vary.*

Find history in Glaoui Kasbah

The once-glorious mountain stronghold of Glaoui Kasbah has been left to crumble into atmospheric ruin, but in the 18th and 19th centuries it was the centre of Telouet's trans-Saharan trading empire. Some 300 artisans worked on its ornate reception rooms, covered with stucco, *zellige* (colourful geometric mosaic tiles) and painted cedar ceilings. *Entry Dh20.*

ATLAS TREKKING TIPS

The Atlas Mountains were made for adventures, and it's easy to explore for nothing in the Moroccan section of the peaks. Be aware, though, that this is serious desert country. For some walks, you'll need a guide from the local Bureau des Guides (around US$40 per day). If trekking independently, carry plenty of water, seek local advice about trail conditions, and respect the weather; be ready for baking heat in summer or sub-zero temperatures in winter.

AFRICA

RED SEA REGION, EGYPT

Accessible from several countries, but most dramatically from Egypt, the Red Sea is a rainbow-coloured underwater wonderland, bordered by deserts. On land or underwater, viewing its marvels can be surprisingly inexpensive.

Dive for less in Hurghada
Less busy and cheaper than Sharm El Sheikh, Hurghada has superb dive sites, including wrecks and Giftun Island Marine Park. Expect close-up views of rays, turtles, nudibranchs, barracuda, moray eels and grey reef sharks. Aquanaut Diving Club has fair rates. *aquanautclub.com; Hurghada Marina; 1-day, 2-dive package €45.*

Free snorkelling
Hippy hangout Dahab is ideal for budget snorkelling, with gear rental available from LE25 per day; hit the coral at Lighthouse Reef and Eel Garden, both in Assalah. The tops of the famous Canyon and Blue Hole dive sites are teeming with life, making for fine snorkelling destinations; half-day boat snorkelling trips cost from LE50.

The Monastery-to-Monastery hike
The Coptic monasteries of St Anthony and St Paul are some of Christianity's oldest monasteries, and hardy trekkers can walk between the two along a 30km trail across the plateau. The 'Devil's Country' should not be underestimated; a local guide is vital so make arrangements in Hurghada. *Free, plus guide fees.*

Trips to St Catherine's Monastery
Visiting the site of the burning bush, where God reputedly spoke to Moses, is a remarkable experience. Travel to Al Milga, the nearest village, by local bus from Suez or Nuweiba, and stay cheaply at El Malga Bedouin Camp (sheikmousa.com; Al Milga; dorm/room from LE30/125). The only fee to visit the magnificent, 1700-year-old St Catherine's Monastery is for the museum. *sinaimonastery.com; free, museum LE80.*

Sinai day hikes
The climb up 2285m Mt Sinai, Egypt's highest peak, has mandatory guide fees ranging from LE175 to LE250. For a free alternative, consider the climb up flat-topped Jebel Mileihis, and the trek along the White Canyon to the Ein Khudra oasis, both accessible from the Nuweiba-St Catherine road. *Free.*

Below: snorkel above the coral reefs near Hurghada, in warm, clear seas.

© Westend61 | Getty Images

VICTORIA FALLS & THE ZAMBEZI, ZAMBIA

There's more to Zambia than its famous river, but you'll want to spend time on this mesmerising waterway, where hippos and crocodiles pop up from the depths. Waterfall trips – including to epic Vic Falls – and river safaris are a splurge, but a small one.

🐘 Budget tips for Victoria Falls

You won't escape the US$20-30 entry fee to see Victoria Falls, so make the investment count. Head to the falls at dawn, when the trail opens, so you see more of the falls and less of other sightseers. Stay cheaply at local legend Jollyboys Backpackers (backpackzambia. com; camping/dorm/room from US$9/12/45) and rent a bike for low-cost pottering around the countryside.

🐘 Big game, small crowds, in Lower Zambezi National Park

It's worth paying the mid-range entry fee for self-drive safaris along this wonderful stretch of the Lower Zambezi River. You'll sight many of the Big Five on the river's fringes; top spots include impala, zebra, buffalo, bushbuck, leopard, lion, cheetah and wild dogs. *lowerzambezi.com; park entry/self-drive vehicle US$25/30.*

🐘 Canoe down the Zambezi

Lodges at all price ranges at Lower Zambezi National Park offer canoe safaris, and nothing beats getting eye-to-eye with a drinking buffalo, or watching dainty bushbuck tiptoe towards the river's edge. Most of the camps and lodges have canoes, so you can go out with a river guide for a few hours from US$50 upwards.

🐘 Witness the Kuomboka Ceremony

This festival marks the ceremonial journey of the Lozi king to his wet-season palace on an elephant-topped barge. The ruler begins the day in traditional dress, but changes into the full uniform of a British admiral; chartered boats offer a chance to join the flotilla, but you may need to camp. *Mar/Apr.*

🐘 Zambia's other falls at Ngonye

As majestic as Victoria Falls but with few other people in sight: welcome to Ngonye. The challenge is getting here; buses run from Sesheke to Sioma, passing near to Ngonye River Camp, which provides cheap digs and offers boat rides for stunning falls views. *ngonyerivercamp.com; Shangombo; camping/chalet from US$10/35.*

Below: you'll sneak up on a lot of wildlife during a serene canoeing trip down the Zambezi River.

AFRICA

© Philip Lee Harvey | Lonely Planet

SERENGETI & MASAI MARA, TANZANIA & KENYA

When it comes to the Big Five, nowhere has it covered like Tanzania's Serengeti and its Kenyan neighbour, the Masai Mara. Both of these world-class reserves have steep ticket prices, but there's plenty of low-cost nature nearby for budget-minded spotters.

Safari on a budget

Serengeti National Park and Masai Mara National Reserve have meaty US$70-80 daily entry fees; your best bet for an economical safari is to book into budget accommodation just outside the park boundaries and make your safari arrangements there. A day-safari starts from US$150 per jeep (with room for six); get a group together and costs fall significantly. *tanzaniaparks.go.tz, kws.go.ke.*

Sleep out under canvas

Both reserves have moderately-priced tented accommodation (and campsites), just outside their boundaries, which see a fair amount of wildlife of their own after dark. In the Serengeti, try Serengeti Stop-Over (info@serengetistopover.com; camping from US$10); in the Masai Mara, try Mara Explorers (maraexplorers.com; camping from US$13.50).

Ukerewe Island, Uganda

A rewarding detour west from the Serengeti, Uganda's Ukerewe Island floats on the edge of Lake Victoria, 50km north of Mwanza in Tanzania. With its rocky terrain broken up by lake vistas and tiny patches of forest, it's a fine antidote to the Serengeti crowds, and you can visit the palace of the island's former king in Bukindo, the main settlement. *Ferry from Mwanza Tsh5000.*

Drive to Lake Natron, Tanzania

The drive from Klein's Gate in the northern Serengeti to Lake Natron has a desolate beauty. The route passes through Masai land, with small *bomas* (fortified compounds) often in view. If you visit from June to November, expect to see some three million flamingos. *Free.*

Relax in Homa Bay, Kenya

What makes Homa relaxing is not the must-see sights, but the lack of safari crowds. From this backwater hub, you can detour inland to the 15th-century Thimlich Ohinga Archaeological Site (entry Ksh1000), or head due east to Ruma National Park, where a US$25 entry fee gives access to rare roan antelopes and Rothschild's giraffes. *Free.*

Below: spot Serengeti gazelles and their predators on a group jeep tour.

© Jonathan Gregson | Lonely Planet

AFRICA

RWANDA, UGANDA & THE DRC

Seeing real-life gorillas in the mist is like stepping into a David Attenborough documentary, but permit costs can be staggering. Be strategic about where you safari, and you'll have money spare for other memorable wildlife encounters.

✋ Cost-effective gorilla spotting

Seeing gorillas is a premium travel experience. Permit fees start at US$1500 in Rwanda and US$600 in Uganda, but just US$400 in the Democratic Republic of the Congo. Transport will add to your costs – consider DRC's Parc National de Kahuzi-Biéga, home to eastern lowland gorillas, and with easy to access from Bukavu.

✋ Waterfall walks in Bwindi Impenetrable National Park, Uganda

Uganda's Bwindi Impenetrable National Park is best known for gorilla safaris, but there's plenty to see for those with smaller travel kitties. Three- to four-hour nature walks run by the Uganda Wildlife Authority (UWA) penetrate the Impenetrable Forest around Buhoma for just US$30, with views of waterfalls, volcanoes and smaller primates. *ugandawildlife.org.*

✋ Primate-spotting hikes in Nyungwe Forest, Rwanda

Nyungwe Forest National Park is Rwanda's most important area of biodiversity. The big drawcard is the chance to track chimpanzees and colobus monkeys, but thrifty travellers can take cheaper hikes along trails, where, in theory, you could run across either primate, along with a slew of birdlife. *visitrwanda.com; guided walks U$40 per person.*

✋ Ride down the White Nile, Uganda

Murchison Falls National Park is Uganda's largest and one of its best, with plenty of animals and waterfalls. Park authorities run three-hour White Nile boat trips from Paraa, promising sightings of hippos, crocodiles and usually elephants. *ugandawildlife.org; entry US$40; boat trip US$30.*

✋ Take the boat to Kisangani, DRC

Kisangani is where the Lualaba River 'becomes' the Congo River, but it feels like a real backwater. The town is ringed by river beaches, but the most amazing thing is the journey here, an epic, even gruelling two- to three-week riverboat ride from Kinshasa that Joseph Conrad would have been proud of. *Ferry tickets start from US$50.*

Below: encountering mountain gorillas, here in Parc National des Volcans, Rwanda, is worth every penny.

AFRICA

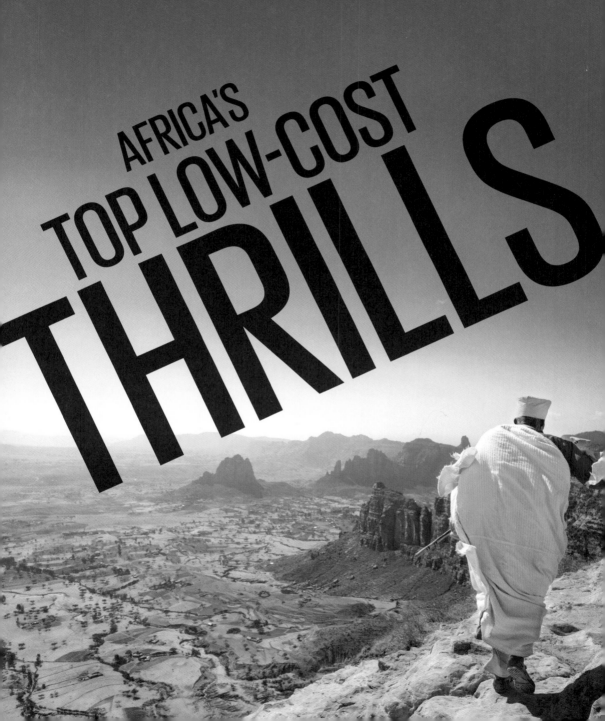

AFRICA'S TOP LOW-COST THRILLS

SURF SWELL IN LIBERIA

Ride some of the best breaks on the continent with a growing cast of Liberian surfers. The golden beaches stretch to the horizon, and at night the crashing waves glow with biolumines- cence. Don't forget to pack your surf- board! *Robertsport, Liberia; free.*

VISIT THE WORLD'S MOST DANGEROUS CHURCH, ETHIOPIA

Carved into a sheer cliff face many centu- ries ago, the church of Abuna Yemata Guh requires visitors to climb a 6m-high vertical wall with no ropes, and negotiate a narrow ledge over a 200m drop. Tips for local guides expect- ed. *Birr150 plus tips.*

DUNE-WALK IN NAMIBIA

While huge swathes of the Namib's famous dunes sit within national parks, those just south of Swakopmund do not. Head out on the road to Walvis Bay and explore. Unique characteristics of the sand mean some dunes hum as you walk along the crests. *Free.*

WILD FLOWERS OF NAMAK- WA, SOUTH AFRICA

The semi-desert of Namakwa erupts in a rainbow of colours each spring when wildflowers carpet the landscape. Let's be clear, this is not a flower show or a larger-scale version of your granny's garden – it is one of Africa's most impres- sive sights. *Aug-Sep; free.*

SET EYES ON THE JADE SEA, KENYA

Lake Turkana's eastern shore at the village of Loyangalani may be at the end of an incredibly long and uncomfortable transport-truck jour- ney, but encounters with Turkana people by the lake described by explorers as the 'Jade Sea' are out of this world. *Lake Tur- kana, Kenya; free.*

MARALAL INTERNATIONAL CAMEL DERBY, KENYA

A carnival atmos- phere pervades at these epic cross-country camel races. Mingle with the crowds, attend the notorious parties and, if you're game, rent a camel and ride in the amateurs' race. *kenyasafari.com/ maralal-camel-derby. html; free.*

HYENAS OF HARAR, ETHIOPIA

The adventurous can take to the narrow alleys of Harar's old town at night to catch glimpses of the city's hyenas. Or shell out Birr100 and you can watch them being fed outside the 16th-century walls each evening – double it and you can feed them yourself. *Free.*

TAKE A DIP IN DEVIL'S POOL, ZAMBIA

Drop into the Zam- bezi River and swim in a natural pool on the very edge of Victoria Falls while water slides past you and plummets over the 100m drop. Free it is not, but it's so wild it is worth every penny (an incredible breakfast is includ- ed). *devilspool.net; Aug-Jan; US$110.*

AFRICA

WELLNESS

Wellness in Africa ranges from the rewarding but rudimentary – wild swims in crystal lakes and basking in natural hot pools surrounded by wilderness – to the lavish and ostentatious, such as massages with crushed grapes and wine baths in the Cape Winelands. Budget travellers should aim to fit in more of the former, fewer of the latter, or see travel costs explode like a champagne cork.

YOGA, MASSAGE & SPIRITUALITY

The African beach scene comes with all

Above: paddle-boarding before Le Morne Brabant in Mauritius. Right: taking a dip in Rochester Falls, Mauritius.

the spa treatments international travellers have grown accustomed to, though in this setting, with the sun beating down on blissful sands, you should feel pretty relaxed already. Cape Town in South Africa is the epicentre of the wellness movement, with inexpensive cafe yoga classes as well as truly extravagant spas in the Winelands.

North Africa has less advertised wellness traditions, with low-cost, male-dominated hammams (traditional Islamic steam baths) found in many historic cities. Beach resorts at sun-n-sand hubs such as Essaouira and Sharm El Sheikh are on hand with more familiar spa experiences, open to men and women, at familiar high prices.

The Indian Ocean is something else again. Asian traditions imported in colonial times meld together with African and European customs, meaning some surprising spiritual experiences, particularly in Mauritius, where Hinduism is the largest religion.

• **Therapy tips** – treatments offered by hotels and guesthouses are often quite pricey, but usually legit; treatments offered on the beach by wandering

practitioners are often cheap, but can be dodgy.

• **Bring your mat** – if you know the poses, you don't need coaching; get up at first light and take your mat to the beach for a (free) DIY dawn yoga session and glow all day.

NATURAL WELLNESS

From the Mediterranean to the Cape, nature lays on free wellness experiences to rival anything on offer at the upmarket spas. Hot springs erupt everywhere, some set in deep bush country, where you can watch antelope amble by while you bask. However, you'll pay a national park entry fee to reach many springs, and some of the best have been developed as health resorts, with predictably high prices.

In the northern deserts, oases offer the opposite experience: cool, crystal-clear desert pools where you can escape the relentless heat. Just remember to seek local advice before entering freshwater in Africa: hazards range from crocodiles to bilharzia and river blindness. In hygiene terms, the South African Cape and Red Sea coasts are probably the safest options.

• **Waterfall swims** – nature's best spa treatment, available for free at hundreds of cascades across Africa; for a fee, bathing in the Devil's Pool on the lip of Victoria Falls (p41) is one of the wildest swims in the world.

• **Wadi bashing** – *wadis* (desert river valleys) offer great wild swimming opportunities across North Africa; just make sure you swim well covered to avoid offending local sensibilities.

• **Eat from nature** – the Indian Ocean offers abundant foraging opportunities. Passion fruit can often be picked off the vine along coastlines, while wild guava can be found on higher ground.

HIGHLIGHTS

MASSAGE ON THE BEACH IN SOUTH AFRICA
A massage on the sand is the vision of tropical beach life, and the coves around Cape Town are a fine spot for a relaxing pummelling at a fair price. Head to Camps Bay or Clifton for a massage to the soothing sound of breaking waves. *p44*

THERMES DE CILAOS, RÉUNION
After trekking the steep mountain trails of the Cirque de Cila-os, you'll have earned some relaxation. Cila-os' *sources thermales* (thermal springs) are heated by volcanic magma, providing a hot, relaxing soak for trail-tired travellers. *p45*

AFRICA

THE CAPE, SOUTH AFRICA

Self-assured Cape Town is the doorway to a world of natural wellness in the stunning scenery of the Cape. Pick from hot springs, natural bathing pools and therapeutic walks surrounded by medicinal plants in the wonderful fynbos that carpets the coastal slopes.

✋ Walk in the fynbos

The fynbos is the Cape's signature form of vegetation, a flower-filled natural shrubbery that includes the rooibos plant, whose leaves are used to make antioxidant-rich redbush tea. You don't need to drink it to feel the benefits; at Silvermine Nature Reserve near Westlake, you can dip in a pool infused with fynbos tannins. *sanparks. org; Ou Kaapse Weg; entry R50.*

✋ Rooftop yoga in Cape Town

Harvest, a stylish cafe and deli in the heart of the Bo-Kaap neighbourhood, is loved for its rooftop views towards Table Mountain and Lion's Head. Come early for morning yoga classes in front of the vista, and a healthy breakfast to look forward to after. *facebook.com/harvestcafect; 102 Wale St; yoga classes from R50.*

✋ Refresh in Avalon's warm pools

Underground spring water gushes from the rock at a balmy 43°C at this hot springs complex. Water is channelled into basking pools with waterslides; it's most peaceful midweek during term time. *avalonsprings.co.za; Uitvlucht St, Montagu; entry R55, R100 Sat & Sun.*

✋ Hike and swim in Crystal Pools

If you're deterred from the sea by fear of sharks, head over to Gordon's Bay and hike to the Crystal Pools of Steenbras Nature Reserve. A moderate trail fee brings access to a series of tannin-coloured pools that are delightful to swim in (or jump into off the rocks). It's a genuine escape from Cape Town's raucous big-city vibe. *capetown.gov.za; Gordon's Bay; entry R75.*

✋ Beach massages at Camps Bay and Clifton

Find a good massage at a budget price on the beach at Clifton and Camps Bay, where Massage on the Beach offers relaxing sessions under a gazebo, starting from R120 for 20 minutes of pummelling in front of the breakers. *massageonthebeach.co.za; Camps Bay & Clifton; Sep-Apr.*

Below: exotic blooms like protea aristata are found across the Cape.

AFRICA

MAURITIUS & RÉUNION

The neighbour islands of Mauritius and Réunion bring a whiff of France to the Indian Ocean, and a wonderful mix of the footloose tropics and old-world sophistication. Beach resorts can be expensive – wellness in nature doesn't have to be.

✋ Find utter peace on the sand

The beaches of Mauritius and Réunion are soundtracked by the gentle swoosh of waves breaking on the reef edge. Swanky resorts have snaffled up some prime stretches, but smaller coastal hubs such as Mauritius' Flic en Flac and Réunion's Plage de Grande Anse have their own stunning sands. Seek quieter beaches down unsigned trails through cane plantations, where all you'll find is the sea, cowries and sand.

✋ Thermes de Cilaos

Cilaos' sources thermales (thermal springs), heated by lava bubbling below the surface, are said to relieve rheumatic pain, among other bone and muscular ailments. Therapies are very affordable; consider a 20-minute hydromassage (€18) – bliss after a long day's hiking in the Cirque. *thermescilaos.re; Route de Bras-Sec, Réunion; treatment costs vary.*

✋ Eat wild guavas, Mauritius

Early April is the prime time to visit Plaine Champagne in Mauritius. The scrub forest that carpets the hills comes alive with locals as people head to the hills to forage for 'goyaves de Chine', small, red, sweet-sour wild guavas that are packed full of antioxidants. Head east from Chamarel to start the harvest. *Free.*

✋ Hike to rippling waterfalls, Réunion

The Indian Ocean's most famous falls are crowded: for a calmer cascade experience, and a natural spot for contemplation, undertake the Haute Mafate trek. On day three you'll emerge from agave scrub at the Trois Roches falls. *Cirque de Mafate; free.*

✋ Visit a branch of the River Ganges

Hindus believe that Mauritius' Grand Bassin is an extension of the River Ganges. As the legend goes, Shiva was circling the globe carrying the sacred river in his hair when a few drops spilled, creating the water-filled crater. Aim to visit during the Maha Shivaratri festival (Feb/Mar; p31), when half a million Hindus make the barefoot pilgrimage to the lake. *Free.*

Below: beach life in Mauritius doesn't need to cost much, once you're there.

AFRICA

THE BEST THINGS IN LIFE IN ASIA: TOOLKIT

For many budget travellers, Asia is the first taste of a wide and wonderful world, with the bonus of bargain prices to make your money go further and your travels last longer. The region isn't as cheap as it used to be, so a bit of savvy planning can go a long way.

Asia's big-ticket things to see – the Taj Mahal, Borobudur, Angkor Wat – have big price tags, and entry fees are often significantly higher for foreigners than for locals. You wouldn't want to miss out on these world wonders entirely, so save on transport, accommodation and organised activities to free up money for these once-in-a-lifetime experiences.

TOP BUDGET DESTINATIONS IN ASIA

Some destinations in Asia are the stuff of budget-travel dreams – here are our top bank-balance-friendly five:

• **INDIA:** The cheapest rooms, meals and train journeys, and more free-to-visit temples than you can count.

• **THAILAND:** The budget gateway to Asia, beloved by generations of backpackers for its low-cost adventures.

• **NEPAL:** Kathmandu can eat through your rupees, but there's nowhere in the world where you can trek so high for so little.

• **LAOS:** Live like a local in this lovely backwater and your money will go far.

• **THE PHILIPPINES:** Asia's wildcard; great for cheap diving and snorkelling and low-cost adventures.

TRANSPORT

Asia is one of the cheapest regions on earth to explore by public transport, but long distances and geographical obstacles – oceans, deserts, mountain ranges – often compel travellers to dig deep for local flights to avoid spending half the trip waiting for trains and buses.

While you can get almost everywhere by public transport, services to outlying

Below: ferries and fishing boats in Aberdeen Harbour, Hong Kong.

© Terry Sze | Shutterstock

villages and sights may only run a few times a week. Unless you have limitless time, it's often worth swallowing the cost of visiting somewhere on an organised tour or by chartered rickshaw or taxi.

AIR

You may prefer to avoid air travel for environmental reasons, but there are times when it's the only option – when getting to Borneo, for example, or when crossing contested borders. Asia has a huge number of discount carriers offering cheap fares online, but check out safety records first. AirAsia (airasia.com) and IndiGo (goindigo.in) have the most extensive networks.

• **Check out the alternatives** – many flights are just a convenient alternative to a cheap overnight boat, bus or train. Plenty of the latter run across international borders, sometimes with a change at the border.

• **Avoid peak seasons** – flight prices surge with demand, which can come from local festivals and religious holidays, as well as from inbound tourists.

• **Get an air pass** – AirAsia's Unlimited Pass (airasia.com) offers unlimited flights on certain routes for a fixed period; the Star Alliance Asia Pass (staralliance.com) covers three to seven flights in a fixed period, but you need to book your flight into Asia on a partner carrier. Similar schemes are offered by Skyteam (Go Asia; skyteam.com) and OneWorld (Visit Asia; oneworld.com).

TRAIN & BUS

Buses and trains are the lifeblood of Asia, transporting millions across vast distances for bargain prices. Indian Railways' enormous network (indianrail. gov.in) covers every corner of the country, for astonishingly low prices.

China, Thailand, Myanmar, Peninsular Malaysia and Vietnam also have efficient, inexpensive networks.

With festivals clogging the cultural calendar, seat availability is one of the biggest considerations. Book early, or you may be pushed into a more expensive class. If all else fails, you can pile into an unreserved carriage and try your luck finding somewhere to sit amongst the melee of people.

• **Bus versus train** – in some Asian countries, trains are cheaper than buses, in others it's the reverse. Trains are almost always more comfortable, and often faster, but departures tend to be less frequent.

• **Travel overnight** – sleeper trains and buses are a great way to cover long distances and save the cost of a night's accommodation. Overnight trains in China and India are a travel experience all by themselves.

• **Show some class** – air-con trains and buses cost more, but could you live with just a fan or an open window? Sleeper class carriages on Indian trains are not significantly less comfortable than air-con carriages, just less chilly.

Above: Japan's Shinkansen trains, here at Kyoto, are a cost-effective way of travelling around the island very quickly.

ASIA

Above: life on the water on Langkawi Island, Malaysia.

• **Look at rail passes** – Indian Railways has stopped its Indrail pass; China's rail pass scheme (actually a prepaid card) doesn't offer fare discounts; the Japan Rail Pass (japanrailpass.net) offers significant savings on Japanese trains, including Shinkansen bullet trains.

BOATS

Boats serve as buses in many parts of Asia, from inter-island sea ferries linking island nations to public boats navigating jungle rivers. Long-distance travel by river and sea can often be surprisingly affordable, though budget airline deals can be cheaper still.

• **Take public boats** – fast tourist boats invariably cost more, so seek out the slow public ferry or the local commuter boat. Overnight boats (including car-ferries) take longer but cost less than fast, passenger-only speedboats.

• **Think safety** – overcrowding is common; lifeboats less so. Flying can often be a safer choice than risking a boat crossing on stormy seas during the monsoon.

LOCAL TRANSPORT

Local buses and urban commuter trains can cost pennies, and even rickshaws and autorickshaws won't eat into your budget much. Taxis (particularly from airports) are another, much more expensive, matter.

• **Share the ride** – Uber, Grab, Ola and dozens of other local rideshare apps open up a world of cheap taxi travel in Asia (though it may be an autorickshaw that actually shows up). Sharing an airport taxi to town with other travellers is another great way to save; ask around at the baggage carousel.

• **Rent or buy a bike** – almost always the cheapest option for local exploring, though cheap rental bikes can be rickety rattletraps. Why not buy? Indian- and Chinese-made bikes are available everywhere at bargain prices.

• **Easy riding** – a rented motorcycle or scooter gives you a much bigger range for local exploring, and prices start from just US$5 per day (a fraction of the cost of renting a car). If you daren't face the roads on your own, charter a car and driver by the day; it's surprisingly affordable, particularly in the Indian subcontinent.

• **Look for passes** – big, modern cities often have transport deals for tourists; Hong Kong's Tourist Passes (mtr.com.hk), Singapore's Tourist Pass (thesingaporetouristpass.com.sg) and Korea's Tour Card (koreatourcard.kr/en) all offer big savings on urban transport.

ACCOMMODATION

Asia has accommodation to suit every budget, and some options that money can only just buy: Maharaja's palaces, five-star penthouses, no-expenses-spared beach resorts. To keep a lid on costs, consider the following:

BUDGET SLEEPING

• **When in doubt, camp** – bringing your own tent will bring down sleeping costs dramatically, particularly in national parks and on mountain treks in India and Nepal.

• **Hit the hostel** – these budget digs often have the best traveller buzz in town to go with the cheap tours, inexpensive cafes and bars and choice of private rooms or dorms. Find beds on hostelworld.com.

• **Institutions mean bargains** – the YMCA (ymca.int), YWCA (worldywca.org), and other local youth travel institutions offer cheap, simple rooms in some great locations.

• **Spiritual stays** – the Indian subcontinent has a rich tradition of ashrams and other religious centres offering rooms for pilgrims and travellers; payment is by donation.

HOTELS & GUESTHOUSES

• **Local versus western** – hotels aimed at tourists are more comfortable (and have better bathrooms) but you pay a mark-up; get used to local-style hotels, shared bathrooms and squat toilets and your money will stretch further.

• **Be a good guest** – small, family-run guesthouses (often in historic shophouses) have great prices, even in the hearts of teeming megacities.

• **Hunt down homestays** – not only do you get an inexpensive room at a homestay (sometimes called *losmen* in Indonesia), you get cultural immersion, and (often) delicious home-cooked food. In remote areas, homestays in tribal villages feel like National Geographic expeditions.

• **Skip the view** – seeing the Taj, or Mt Fuji, or the ocean from your room is a costly extra that doesn't always justify the hike in room rate.

• **Ask for a discount** – you never know, and many places have so much competition from rival hotels that they'll consider it.

RESORTS

• **Stay near, not beside, the beach** – beachfront hotels invariably charge more for sand on demand; you'll find cheaper rooms without direct beach access just inland.

• **See the bigger picture** – resorts will encourage you to do everything on site, but there may well be cheaper choices elsewhere – there's no obligation to eat or book tours in the place where you're staying.

• **Be season savvy** – resorts always charge more when the weather is best, when the views are clearest, and or when the festival is in full swing; travel off-season and room rates plummet.

ASIA

TERRIFIC TEMPLES

AT TINY PRICES

Asia's top temples and ancient sites come with towering entry fees, but thousands of other temples, ruins, stupas, mosques, monasteries and shrines serve up spirituality for free.

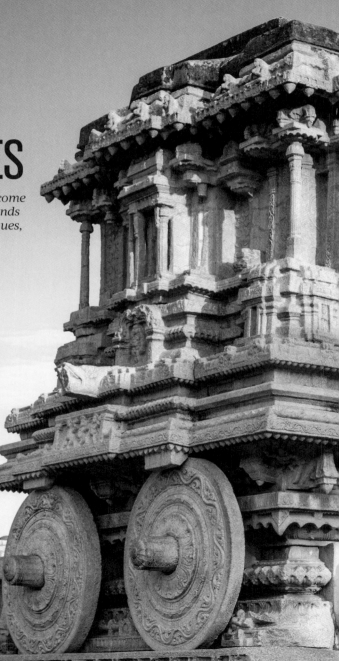

WAT PHU, LAOS

Tucked away near the banks of the Mekong River, the sprawling, Khmer-era ruins of Wat Phu sit in graceful decrepitude. The location, on stepped terraces above the river, is dramatic, and the general lack of crowds and dramatic natural setting evoke a uniquely soulful response. *Near Pakse, Laos; entry 50,000K.*

HAMPI, INDIA

Rather amazingly, there's no entry fee to visit most of the ruined temples, palaces and shrines at Hampi, ancient capital of the Vijayanagar Empire. Make time for the Hanuman Temple at Anegundi, the imposing Achyutaraya Temple at Sule Bazaar and the still-used Virupaksha Temple at Hampi Bazaar. *karnataka.com/hampi; Hampi, Karnataka; free.*

MASJED-E NASIR AL MOLK, IRAN

One of the most elegant buildings in Iran, Shiraz' 'Pink Mosque' was built at the end of the 19th century and its coloured tiling (an unusually deep shade of blue) is the epitome of Islamic grandeur. The stained glass, carved pillars and polychrome faience of the winter prayer hall dazzle the eye when the sun streams in. *nasiralmulk.ir; off Lotf Ali Khan Blvd, Shiraz; entry IR150,000.*

BANTEAY CHHMAR, CAMBODIA

Admission to the temples at Angkor Wat will take a ravenous bite out of your travel funds, but there's no such problem at Banteay Chhmar. Beautiful, peaceful and covered in astonishingly intricate bas-reliefs, this 12th-century Khmer masterpiece combines temple grandeur with inexpensive homestay accommodation. *visitbanteaychhmar. org; Banteay Chhmar village; entry US$5.*

CHANGU NARAYAN TEMPLE, NEPAL

The regal Durbar Squares in Kathmandu, Patan and Bhaktapur come with a similarly regal entry fee, but some of the Kathmandu Valley's grandest temples are free. Crowning a hilltop east of Kathmandu, the Unesco-listed Changu Narayan Temple is a living museum of carvings from the Licchavi period, rising in two handsome tiers in an ancient courtyard. *changunarayan.org. np; Changu; free.*

Left: the Stone Chariot in Hampi's Vittala temple complex.

LAMA TEMPLE, CHINA

Once the most revered Tibetan Buddhist temple outside Tibet, Beijing's Lama Temple was originally the royal residence of the prince who would become the Yongzheng Emperor. For a modest entry fee, you can wander through halls and courtyards, past prayer wheels and clouds of incense smoke. *yonghegong. cn; 12 Yonghegong Dajie, Beijing; ¥25*

PRASAT PHIMAI, THAILAND

Off the main tourist trail (so entry fees and accommodation are easy on the pocket), the ruined temple of Prasat Phimai was built as a Mahayana Buddhist temple in the 10th century, but it's dotted with carvings of Hindu deities and design elements that were later used at Angkor Wat. *Th Ananthajinda, Phimai; entry 100B.*

ARTS & CULTURE

Asia doesn't just have sights, Asia is a sight. Simply walking down a back street or driving through the lush, green countryside is a vivid, supersaturated spectacle. When you dip into the continent's towering temples, ruined civilisations, fabulous festivals and traditional art, the colour and energy is turned up to 11.

This makes Asia extremely cost effective for modern-day explorers. When every encounter is an experience, and every moment is a photo opportunity, you don't need to feel bad about missing

Above: a Buddhist temple in Beijing. Right: the fantastic Gyeongbokgung Palace in Seoul.

a few museums, galleries or famous monuments.

ASIA'S TOP SWITCHEROOS

Everyone wants to see Asia's world wonders, but steep entry fees can diminish some of the wow. For cheaper alternatives; sample some of the following:
• **SKIP:** The Taj Mahal, Agra, India (entry US$14.20)
• **SEE:** Jama Masjid, Old Delhi, India (free)
• **SKIP:** Taktshang Goemba, Paro Valley, Bhutan (daily tourist fee US$200-250)
• **SEE:** Thiksey Gompa, Ladakh, India (free)
• **SKIP:** Angkor Wat, Siem Reap, Cambodia (entry from US$37)
• **SEE:** Wat Phou, Champasak, Laos (entry US$5.40)
• **SKIP:** Borobudur Temple, Java, Indonesia (entry US$25)
• **SEE:** Koh Ker, Siem Reap, Cambodia (entry UDS$10)
• **SKIP:** The Forbidden Palace, Beijing (entry US$9)
• **SEE:** Gyeongbokgung Palace, Seoul, South Korea (entry US$2.70)

ASIA'S CULTURE FOR LESS

The big sights in Asia often have two-tier entry fees – a cheap price for locals, and a higher US dollar fee for foreigners. This ensures that attractions aren't just accessible to (often comparatively wealthy) tourists; and the fees keep lots of local people employed. To save on sightseeing, prioritise religious sights (they often don't charge) and landmarks you can admire free from outside – in Asia, you'll be spoilt for choice.

• **Respectful visitors** can enter many active places of worship for free, from mosques to Hindu temples, but a small donation is usually appropriate. You may be asked for elevated sums as a foreigner, though amounts are still relatively low.
• **Ruinous ruins** – ruins almost always have entry fees, but prices drop the further you get from the tourist trail. Roam off the beaten track everywhere from India to Kazakhstan and you'll find lost civilisations at spare-change prices.
• **Tips for institutions** – few museums are free and few have free days, but admission can be well worth paying considering the treasures on display. There's often a charge to take photos or video; you'll save if you leave your camera in your hotel room.
• **Museum streets** – in places like Kathmandu and Hampi you can wander for free around temple carvings and statuary that would be locked behind glass anywhere else in the world.
• **Seek out tourist passes** – the Singapore Tourist Pass is one good deal (thesingaporetouristpass.com.sg). Youth travel operators such as Intrepid (intrepidtravel.com) offer discounts if you take multiple tours in the region.
• **Guide fees** – sometimes the free map that comes with your ticket is enough; where you do need a guide, ask at guesthouses for recommendations.

HIGHLIGHTS

DELHI'S CHANDNI CHOWK
Delhi's most atmospheric thoroughfare has been crammed with traders since medieval times, and it offers time travel to an earlier age. Take a free wander past market vendors, ancient temples and food stands that have been there for generations. *p55*

WAT ARUN, BANGKOK
The small entry fee at this sublime riverside temple buys a lot of history and atmosphere. Wat Arun's towers are covered in mosaics made from ancient shards of broken porcelain, and towering demon-giants guard its gateways. *p56*

ASIA

DELHI, INDIA

*Delhi offers a seductive glimpse of India in miniature. Many of its top experiences –
touring temples, stumbling over Mughal monuments in the backstreets, diving into
bazaars – are free, and there are regular opportunities to pause for a cheap cup of chai.*

India Gate & the Secretariat

India Gate is the heart of the former British colonial capital, an imposing 42m-high stone memorial arch designed by Lutyens in 1921 to pay tribute to India's war dead. From here, Rajpath cuts west to the Rashtrapati Bhavan (President's House) and the grand government buildings of the Secretariat. *Rajpath; free.*

Agrasen ki Baoli

This atmospheric 14th-century step-well was set in the countryside until the city grew up around it. From street level, 103 steps descend to the bottom, flanked by arched niches. It's a remarkable discovery among the office towers southeast of Connaught Place. *Hailey Lane; free.*

Jama Masjid

India's most magnificent mosque was a Mughal creation, concocted from white marble and red sandstone during the reign of Emperor Shah Jahan; its courtyard has room for 25,000 of the faithful. Entry is free, if you leave your camera behind and don't climb the minaret (but even that won't break the bank). *Urdu Bazar Rd, Old Delhi; free, minaret entry Rs 100.*

Lodi Gardens

People-watching is the focus of this lovely green lung: join laughing yogis, canoodling couples, dog-walkers, runners, elderly strollers, casual cricketers or family picnickers in this idyllic park and see what makes Delhi tick. All activity takes place against a fabulous backdrop of landscaped gardens, crumbling tombs and Mughal monuments. *Enter at South End Rd; free.*

Gurdwara Sahib Bangla

This magnificent white-marble gurdwara (Sikh temple), topped by glinting golden onion domes, was constructed at the site where the eighth Sikh guru, Harkrishan Dev, stayed before his 1664 death, tending to victims of Delhi's cholera and smallpox epidemic. The waters of the gurdwara's tank are said to have healing powers. *dsgmc.in; Ashoka Rd; free.*

Below: the Hindu temple of Akshardham features a musical fountain that relates a story from the Kena Upanishad.

VISV

Hauz Khas

Built by Sultan Ala-ud-din Khilji in the 13th century, Hauz Khas means 'noble tank', and the reservoir at the centre of the ruins is dominated by the impressive ruins of Feroz Shah's 14th-century *madrasa* (religious school) and tomb, and more mausoleums spill through the surrounding streets and the adjacent deer park. *Hauz Khas; free.*

Gandhi Smriti

This poignant memorial to Mahatma Gandhi is in Birla House, where Gandhi was staying when he was shot dead on the grounds by a Hindu zealot on 30 January 1948, after campaigning against intercommunal violence. The exhibits include film footage and rooms preserved just as Gandhi left them. *5 Tees Jan Marg; closed Mon and 2nd Sat of month, free.*

Crafts Museum

Not quite free but offering maximum bang for your rupee, this lovely museum sprawls over a compound full of tree-shaded carvings and examples of village huts from India's regions . Displays celebrate the traditional crafts of India's myriad states, with some beautiful textiles on display indoors. *Bhairon Marg; closed Mon; Rs 200.*

Indira Gandhi Memorial Museum

In the residence of former prime minister Indira Gandhi, this fascinating museum explores the lives of several generations of her political-heavyweight family. Exhibits include the sari the former PM was wearing when she was assassinated in 1984. *1 Safdarjang Rd; closed Mon; free*

Akshardham Temple

Delhi's largest temple, the Hindu Swaminarayan Group's Akshardham is breathtakingly lavish. Artisans used ancient techniques to carve the pale red sandstone into elaborate reliefs, including 20,000 deities, saints and mythical creatures. *akshardham.com; National Hwy 24; free*

Sunder Nursery

The modest entry fee for this restored Mughal garden is a small price to pay for stepping back into the grandeur of Mughal times. Clipped lawns, delicate waterways and a network of paths dotted with fruit trees, flower beds and tree-shaded benches connect 16th-century Mughal tombs and pavilions. *Mathura Rd; Rs 100.*

Hazrat Nizam-ud-din Dargah

Visiting the marble shrine of Muslim Sufi saint Nizam-ud-din Auliya is Delhi's most mystical, magical experience. The *dargah* is hidden away in a tangle of bazaars selling rose petals, *ittars* (perfumes) and offerings, and on some evenings you can hear *qawwalis* (Sufi devotional singing). *Off Lodi Rd, Nizamuddin Basti; by donation.*

INTACH Heritage Walks

Led by passionate historians from INTACH (Indian National Trust for Art and Cultural Heritage), these weekend walks are a good way to see the city through the eyes of its residents, including sights such as Mehrauli Archaeological Park. Online registration required. *intachdelhichapter.org/heritage_walks.php; various locations; Rs 200.*

Evenings at India International Centre (IIC)

Set in leafy surrounds, the IIC buzzes with energy as academic old-timers rub shoulders with trendy young hipsters, all here for the gigs, performances and debates. Most events are open to the public, and there's also a massive library. *iicdelhi.nic.in; 40 Max Mueller Marg; free.*

BROWSE OLD DELHI'S BAZAARS

The tangle of lanes between Chandni Chowk and Paharganj is one of the world's most evocative districts. Whole streets in this bewildering warren are devoted to the sale of wedding outfits, kites, stainless steel pots, saris, wrapping paper, spices and more. Getting agreeably lost in the maze is like stepping into a magical world, and some of Delhi's best street-food stalls are hidden away in the old bazaars.

ASIA

BANGKOK, THAILAND

*All that glitters is gold in Bangkok, where every turn reveals another gleaming Buddha.
You don't need a princely budget to see Bangkok's cultural treasures; in Thailand's City of
Angels, you can experience Asia's timeless rhythms with just a pocket full of change.*

✋ Wat Arun

The small fee at this Bangkok temple
delivers a lot of wonder. Approached from
the river, the towering *prangs* (Khmer-
style towers) of the Temple of Dawn look
like carved stone; up close, they reveal
themselves as gardens of mosaic flowers
in broken Chinese porcelain. Dip your
head respectfully to the *yaksha* (guardian
giants) as you enter the statue-filled
gardens. *watarun1.com/th; off Th Arun
Amarin; entry 50B.*

✋ Golden Mount (Wat Saket)

The low-rise Ko Ratanakosin and
Banglamphu districts are big on top
sights but short on viewpoints from
where they can be seen. That's where
the Golden Mount comes in. Just east of
the Democracy Monument, this artificial
mountain is crowned by a looming golden
chedi (stupa) and the terrace offers bird's-
eye city views. *Off Th Boriphat; free.*

✋ Ban Baat

Thailand's monks need freight-loads
of begging bowls (*bàht*) for collecting
alms, and this tiny urban village keeps
the traditional bowl-making art alive.
Jump off the *khlong* boat at Tha Phan
Fah and you'll hear the tap, tap, tap
of metalworkers hammering bowls
together from eight leaves of different

metals, symbolising the eightfold path of
Buddhism. *Soi Ban Baat; free.*

✋ Baan Silapin Artists' House

Hidden away on a Thonburi *khlong* (canal),
this historic wooden house is home to
traditional puppeteers, training the next
generation in this ancient Southeast

Below: Wat Arun is
known as the Temple
of Dawn, but it's
just as stunning at
sunset.

© Jirawas Teekayu | 500px

ASIA

Asian art form. Afternoons from 2pm, the black-clad puppetmasters bring Thai legends to life; it's highly nostalgic and very atmospheric. *Wat Thong Sala Ngam, Soi 28, Th Phet Kasem; closed Wed; free.*

Bangkokian Museum

The pocket-sized Bangkokian is an oft-missed jewel in a city where most treasures are proudly on display. Spread over a series of dainty homes are rooms full of personal belongings that look as if the owners stepped out the front door to pick up some noodles and never came back. *facebook.com/BkkMuseum; 273 Soi 43, Th Charoen Krung; closed Mon; free.*

Amulet Market

Thai Buddhists have been trading religious amulets for centuries, and collectors congregate at the amulet-stacked stands that dot the pavement along Th Maha Rat, haggling over particularly sought-after Buddhist tablets, votive objects and oversized phallic symbols. Look out for committed aficionados, struggling to stand under the weight of all their amulets. *Th Maha Rat; free.*

Ko Kret Island

Bangkok's tiny Burmese enclave feels like a country village, smuggled into the northern suburbs and secreted inside a loop of the Chao Phraya River. Founded in 1722, this miniature community of potters follows an 18th-century pace of life. Wander boardwalk lanes past potters' studios to a tiny, wonky riverside *chedi* in the Mon style. *Nonthaburi; free.*

Chao Phraya Express ferry

For about the price of a bowl of rice, you can board Bangkok's most atmospheric vantage point on a budget river cruise. The best time to ride the ferry is at dusk;

join the monks waiting at Tha Phra Athit pier and cruise past some of Krung Thep's (Bangkok's) landmark monuments. *chaophrayaexpressboat.com; Th Phra Athit Pier; 15-50B.*

Lumphini Park

Before Bangkok erupts into life each morning, a gentler awakening takes place in Lumphini Park, where old-timers gather to practise tai chi before an audience of monitor lizards and the odd jogger. For a few precious hours the city seems to float in a bubble of calm. *Free.*

Pak Khlong Talat Flower Market

Locals swing by this atmospheric market to sweeten the air with perfumed blooms as part of their daily routine, and a hint of jasmine scents every Bangkok taxi. At festivals and celebrations, though, the flower arrangements on offer go into overdrive, with outrageous floats of flowers and banana leaves for processions and home shrines. *Th Chakraphet; free.*

Student bands in Banglamphu

Every evening, singer-songwriters and teen bands take to the stage in the hip little bars around Soi Rambutri, just a street over from the tourist-oriented antics on Thanon Khao San. Many are students from nearby Thammasat University, and there's a lot of talent on display. *Free for the price of a beer or snack.*

People watch at Chatuchak

Bangkok's legendary weekend market is a bazaar to end all bazaars, selling everything from chopsticks to spirit houses and jeans to live goldfish. Bargains abound, but just wandering and people-watching around the thousands of stalls is one of Bangkok's greatest spectacles. *587/10 Th Phahonyothin, Sat & Sun; free.*

STAYING REFRESHED IN BANGKOK

Exploring Bangkok can be thirsty work, but a bottle of Chang beer can cost more than a meal, and wine and spirits can empty your wallet in a heartbeat. Follow the lead of locals and stock up on refreshments at the nearest 7-Eleven, saving posh sit-down drinks for special occasions. Your hotel balcony is the best spot for an impromptu drinks party; public drinking is frowned upon, and banned in parks and religious sites, including the grounds of monasteries.

ASIA

KYOTO, OSAKA & NARA, JAPAN

Bullet trains whooshing between Kyoto and Osaka are impressive, but the cultural wonders that you encounter when you disembark will blow your mind. Save your cents visiting temples, museums and markets – you'll need them for sleep stops and meals!

Arashiyama Bamboo Grove, Kyoto

With green stalks in every direction, this famous bamboo grove is like stepping into a historic woodcut. It's most atmospheric on the approach to Ōkōchi Sansō villa, garden-ringed home of Samurai-movie actor Ōkōchi Denjirō. *kyoto-bamboo.jp; Ogurayama, Saga, Ukyō-ku; free*

Fushimi Inari-Taisha, Kyoto

With arcades of *torii* (shrine gates) spread across a mountain, this shrine complex is a world unto itself. The access pathway climbs the slopes of Inari-san, lined with atmospheric sub-shrines. *inari.jp; 68 Yabunouchi-chō, Fukakusa, Fushimi-ku; free*

Tōfuku-ji, Kyoto

Home to a garden dotted with pavilions, Tōfuku-ji is the living image of Japanese spirituality. This Buddhist temple complex includes 24 sub-temples, and the gardens offer views to the Tsūten-kyō (Bridge to Heaven). *tofukuji.jp; 15-778 Honmahi, Higashiyama-ku; free, gardens ¥400*

Kasuga Taisha, Nara

Founded in the 8th century, this lantern-filled shrine at the foot of Mikasa-yama peak was created to protect the new capital, Nara. It was ritually rebuilt every 20 years until the late 19th century. *kasugataisha.or.jp; 160 Kasugano-chō; free*

Abeno Harukas, Osaka

The observatory on the 16th floor of Japan's tallest building (300m, 60 storeys) is free, but you may be tempted to pay for the top-level Harukas 300 observation deck .Both offer a bird's eye view of Kansai's streets. *abenoharukas-300.jp; 1-1-43 Abeno-suji, Abeno-ku; free; observation deck ¥1500*

National Museum of Ethnology, Osaka

This museum displays everything from traditional masks to Ghanaian barbershop signs, Bollywood movie posters and even a Filipino jeepney bus. Don't miss the music room, where you can summon global street performances via a touchpanel. *www.minpaku.ac.jp/english; 10-1 Senri Expo Park, Suita; closed Wed; free.*

Below: a *torii* gate at Fushimi Inari-Taisha shrine in Kyoto, marking the transition from the earthly to the sacred.

© Takashi Images | Shutterstock

ASIA

BEIJING & THE GREAT WALL, CHINA

It was once almost free to explore Beijing's ancient core, but prices are climbing, and most visitors splurge a few dollars to explore the Forbidden City and Summer Palace. Balance the budget with what you can see for free, including Beijing's medieval hútòngs.

798 Art District
Beijing's best contemporary art galleries bloom in the concrete skeleton of a former factory. After wandering the Bauhaus-style buildings, swing by Beijing Tokyo Art Projects (tokyo-gallery.com; free) and UCCA (ucca.org.cn; free Thu). *Dà Shānzi; cnr Jiuxianqiao Lu & Jiuxianqiao Beilu; free.*

Beijing Museum of Natural History
This collection covers everything from creepy crawlies to prehistoric monsters. The star of the dinosaur exhibit is Mamenchisaurus jingyanensis, a 26m-long lizard that stomped around China in the Late Jurassic. Book a day in advance. *bmnh.org.cn; 126 Tianqiao Nandajie; free.*

Hútòng hiking
Beijing's beating heart can still be felt in the *hútòngs* – a maze of laneways, surrounding the Drum Tower, and sprawling around Dashilan. Or, check out the imperial *hútòng* courtyards, once home to royal eunuchs, northeast of Jingshan *Park. 24hr; free.*

National Museum of China
Delve into 5000 years of state-approved (and selective) Chinese history, culture and art, in exhibitions covering everything from prehistory to China's descent into semi-feudalism after the 1839-42 Opium War, and the march to 'national happiness and prosperity'. *en.chnmuseum.cn; 16 East Chang'an Ave; closed Mon; free.*

Beihai Park
With a history spanning five dynasties, this island park boasts temples, pavilions and lake views. You can view the waterlily-filled ponds of the 'Quieting Heart Room' – where emperors retreated for reflection and relaxation. *beihaipark.com.cn; 1 Wenjin St, Xichéng; admission high/low season ¥10/5.*

Jiankou Great Wall
This wild, unrestored section of the Great Wall is free to visit, and powerfully evocative. There's a small charge (¥20) to enter historic Xizhazi village, the nearest settlement. *Jiankou, Xizhazi, Huairou District; free.*

Below: kilometres of battlements along the Great Wall of China stretch into the distance.

ASIA

FOOD & DRINK

As every traveller to Asia knows, the best eats are out on the street. Every bus stand, marketplace, beachfront and town square buzzes with stalls selling street-food bargains. By skipping sit-down restaurants and grazing al fresco, you can reduce your eating costs to almost negligible levels – assuming you can resist the temptation to eat at every stall you pass by.

STREET FOOD

You don't have to look far to find a cheap meal on the street. Stalls fill

Above: shop for fresh veg in Hoi An, Vietnam. Right: Kyoto's Nishiki market. Overleaf: street cuisine in Bangkok's Chinatown.

every marketplace, waterfront, transport hub, drinking district and business quarter. And Asia's night markets are an institution – street corners fill up with mobile food carts every afternoon and throng with diners until long after dark. However, there's an art to street feasting. Here are some tips:

• **Stick to busy stalls** – to avoid stomach bugs, look for the stands crowded with locals; empty places are usually empty for a reason.

• **Seek out local delicacies** – some towns have become culinary legends by virtue of a single dish, served at tiny prices at local food stalls.

• **Time it right** – the best food markets in Asia take place at night, or more accurately, from mid-afternoon till the early hours of the morning.

• **Go where the commuters go** – the cheapest meals in town are often served at bus and train stations or in the lanes between office buildings, where workers come to chow down.

• **Markets mean munchies** – every produce market has inexpensive street-food stalls and even cheaper places to buy a meal pre-bagged or boxed to take away.

SIT-DOWN DINING

Restaurant eating will raise your travel costs, but can also take you to gastronomic heaven. The golden rule for budget eating is go where the locals go; places aimed at tourists charge higher prices (usually for less tasty food).

Set aside your squeamishness about unusual ingredients and you'll find Asia's finest flavours at frugal prices. For less brave palates, cafes in traveller centres serve a dizzying medley of international dishes – pizza, pasta, burgers, tacos, kebabs – at prices that won't leave a big dent in your wallet.

• **Seek out restaurants** crammed with locals: plastic chairs, minimal decor and hordes of diners are usually good signs.

• **Eat where you stay** – most guesthouses and hotels have an in-house restaurant serving economic meals for guests; they're rarely the best meal in town, but rarely the worst.

• **Know the menu** – some dishes are just downright expensive (hello ginseng chicken), but there are often cheaper alternatives that are just as tasty.

• **Sides are main courses** – dishes that you might think of as a side-serving (fried rice, stir-fried greens, soups) can be a meal all by themselves.

• **Budget breakfasts** – a bowl of *congee* (rice porridge), a *dosa* (rice and lentil pancake) or a serving of *roti canai* (flatbread with spicy sauce) is vastly cheaper than a Western breakfast.

HIGHLIGHTS

TIM HO WAN, HONG KONG, CHINA

Dim sum comes with a Michelin star at this small chain of Hong Kong eateries, owned by a former Four Seasons dim sum chef. It's a great place to get the hang of ordering the delectable small breakfast plates that make up Hong Kong's most famous feast. *p66*

HAWKER CHAN HONG KONG SOYA SAUCE CHICKEN RICE & NOODLE, SINGAPORE

It will take you longer to say the name than to eat at this Michelin-starred food stall, where Mr Chan Hon Meng serves up legendary cara-melised pork char siew and soya-sauce chicken. Waiting times can reach two hours, but it's worth every second. *p67*

ASIA

行
ทองสุพรรณ
ทองสุพรรณ
THONG SUPHAN YAOWARAJ 行金大興發成
海外天魚翅酒樓
ไม่มีสาขา

• **Be smart about seafood** – some fish and crustaceans cost more than others, so pick wisely; eat close to where fish are landed for more savings.

• **Account for taxes** – the more formal the setting, the higher the chances that expensive taxes will be added to your meal.

PLATE MEALS

Asia is the original home of the plate meal – an all-you-can-eat feast of meat and/or veg in sauces, served with bottomless rice or flatbreads – the Indian *thali* is the most famous incarnation. This is the the most bargainous way to eat at lunchtime, and spreads as cheap as US$0.30 are not unheard of.

Nepal's *dhal bhaat* is a less ambitious riff on the same idea (typically, a veg curry, lentil soup and bottomless rice), while Sri Lanka's 'curry and rice' is the island equivalent, expanded to include fish curries, dry fried dishes and *hoppers* (either lentil-and-rice pancakes or tangles of noodles). Middle Eastern *mezze* spreads have a similar sampling-plate theme.

In Southeast Asia, look for local restaurants with a line of big metal pots or steel food trays out front. Ordering is a case of pointing at whatever looks tasty: the food is freshly prepared and prices are tiny. In Malaysia and Indonesia, track down *nasi kendar* restaurants run by Indian Malaysians, serving buffet-style spreads of rice and curries at bargain prices.

RELIGIOUS & FESTIVAL FOOD

Festivals are a special time to eat in Asia. Every celebration comes with its own delicacies, sold at snacking prices whenever there's a festival in town. Sweet treats are favourites, but there are plenty of savoury snacks too.

Some Southeast Asian temples lay on free food for devotees (and visitors) on feast days. In India, many Krishna temples offer free (or discount) vegetarian meals to pilgrims. And Sikh gurdwaras invariably have a *langar* (communal kitchen) where all-you-can-eat meals are dished out for free to all comers (a donation is appropriate), sometimes thousands of plates at a time.

SMART DRINKING

Drinking is an area where it can be hard to keep a lid on expenses. Most countries in Asia impose high alcohol taxes, and some restrict the availability of alcohol for religious reasons, boosting prices.

• **Buy takeout drinks** at convenience stores (such as 7-Elevens) and liquor marts to drink in your hotel or on the beach.

• **Respect religion** – Muslim countries, and some Hindu and Buddhist areas, make it difficult to obtain alcohol, raising prices. Learn to go without, drinking teas, coffees or juices instead, and you can be respectful while saving a fortune.

• **When to wine** – wine is invariably expensive – if wine is your poison, seek out local producers such as Thailand's Monsoon Valley or India's Sula and Grover.

• **Lubricate dinner** – drinking at restaurants is almost always cheaper than at bars and clubs.

HIGHLIGHTS

BADEMIYA SEEKH KEBAB STALL, INDIA
Diners drive across Mumbai to sample the moist, spicy kebabs at this Colaba hole-in-the-wall, turning every night into a tailgate party. It's a meat feast with few frills – most eat in their cars or on the curb – but few leave unsatisfied. *p66*

ANJUNG GURNEY NIGHT MARKET, MALAYSIA
Penang's hawker courts are living legends, and this is the biggest and best loved, serving everything from flame-cooked satay sticks to fried oysters and *loh bak* (five-spice pork sausage). Come early and graze from stall to stall... *p68*

ASIA

1 WARUNG BABI GULING, UBUD, BALI

If you thought suckling pig was just for medieval monarchs, think again. In Indonesia, roast *babi guling* is food for the people, and Warung serves roast-pork plates for rupiah-saving prices. *Jl Tegal Sari, Ubud, Bali; mains from 55,000Rp.*

2 MAVALLI TIFFIN ROOMS, BENGALURU, INDIA

A legendary name in South Indian comfort food, this eatery has had Bengaluru eating out of its hand since 1924. Head to the dining room upstairs for delicious *idli* (fermented rice cakes) and *dosa* (savoury crepes) served with silverware. *mavallitiffinrooms.com; 14 Lalbagh Rd, Mavalli, Bengaluru; closed Mon; meals from Rs 90.*

3 TALAT CHANG PHEUAK NIGHT MARKET, CHIANG MAI, THAILAND

Chiang Mai gourmands head straight for the night markets. At Talat Chang Pheuak, the Cowboy Lady's stall serves the finest *kôw kăh mŏo* (stewed pork leg with rice) in the north. *Th Mani Nopharat, Chiang Mai; dishes from 30B.*

4 CRAB MARKET, KEP, CAMBODIA

Famed for its fat, sweet mud crabs, Kep-sur-Mer is still the best place to feast on these cracking crustaceans. At Kep's crab market, crabs fried with Kampot pepper are a taste sensation – follow the crowds to the best of the waterfront restaurants. *Kep; 1kg crab from 50,000r.*

ASIA'S BEST CHEAP GOURMET GRUB

You don't need big bucks to eat like a king in Asia: some of the finest food is served for discount prices on streets across the continent.

5 DAIWA SUSHI, TOKYO, JAPAN

Sampling sushi at source is de rigueur on a trip to Tokyo, and sashimi is sliced straight from the day's catch at Daiwa Sushi, where set-menu lunches are ocean-fresh and (by Tokyo standards) friendly on the pocket. *1st fl, Bldg 5, Toyosu Market, Kōtō-ku; closed Sun; set meals from ¥4400.*

6 LIN HEUNG TEA HOUSE – HONG KONG, CHINA

Finding discount dim sum is a challenge in Hong Kong, but Wellington St's Lin Heung serves up a bargain feast: come for steamers on trolleys, sulky service and delicious, inexpensive morning dim sum that makes it all worthwhile. *160-164 Wellington St, Hong Kong; dim sum from HK$15.*

8 SERAY-E MEHR TEAHOUSE, SHIRAZ, IRAN

This old-school Iranian teahouse is an initiation into a vanishing way of life. Hidden behind a small door next to the Seray-e Moshir caravanserai, this museum-piece has a small menu of tasty favourites – *dizi* (mutton stew), *kubideh* (kebabs), *zereshk polo* (Persian chicken) – plus tea, of course, at bargain prices. *Seray-e Mehr, Bazar-e Vakil, Shiraz; meals from IR250,000.*

7 NASI KANDAR PELITA, KUALA LUMPUR, MALAYSIA

Malaysia gets a double dip in this list thanks to its fabulous Mamak hawker food; at 24hr Nasi Kandar Pelita, a meal of exquisite *roti canai* (fried flatbread with daal and chicken curry) will set you back mere pennies. *149 Jln Ampang, Kuala Lumpur, RM2-8*

HONG KONG, CHINA

Hong Kong has a reputation as the big-bucks stopover in Asia, and it's easy to burn through the renminbi feasting on dim sum for breakfast and bar-hopping by night. But with imagination, you can still feast and party without a banker's budget.

Yat Lok, Central
Bump elbows with locals at this basic joint known for its roast goose. Anthony Bourdain gushed over the bird. The leg is the most prized cut, and the general rule is the more you pay, the better your meat will be. *34-38 Stanley St, Central; meals from HK$56.*

Tai Cheong Bakery, Central
Tai Cheong was best known for its beignets (*sa yung* in Cantonese) until former governor Chris Patten was photographed wolfing down its egg-custard tarts. Since then, 'Fat Patten' egg tarts have hogged the limelight. Frankly, it's all delicious. *35 Lyndhurst Tce, Central; pastries from HK$6*

Lan Fong Yuen, Central
The rickety facade hides a *cha chaan tang* (tea cafe). Lan Fong Yuen is believed to be the inventor of 'pantyhose' milk tea (poured through a stocking-like filter), and droves of Instagrammers come to worship here. Over a thousand cups of the strong and silky brew are sold daily alongside pork-chop buns and other hasty tasties. *2 & 4A Gage St, Central; closed Sun; meals from HK$60.*

Kau Kee Restaurant, Central
You can argue till the noodles go soggy about whether Kau Kee has the best beef brisket in town. Whatever the verdict, the meat – served in a fragrant broth – is hard to beat. During the 90 years of the shop's existence, film stars and politicians have joined the queue for a table. *21 Gough St, Central; closed Sun; meals from HK$50.*

Tim Ho Wan, Kowloon
Renowned as the world's first budget outlet to receive a Michelin star, Tim Ho Wan has spread from its Mong Kok roots into a mini empire. There are now five restaurants in Hong Kong, including this one, serving dim sum plates. *timhowan.com.hk; 9-11 Fuk Wing St, Sham Shui Po; dim sum from HK$28.*

Ap Lei Chau Market Cooked Food Centre, Ap Lei Chau (Aberdeen Island)
Above an indoor market, *dai pai dong* (food stall) operators cook up a storm in a hall littered with folding tables and plastic chairs. Pak Kee and Chu Kee offer simple but tasty seafood dishes; or bring fresh seafood from the market to be cooked as you like. *Ap Lei Chau Municipal Services Bldg, 8 Hung Shing St, Ap Lei Chau; mains from HK$45.*

LockCha Tea Shop, Wan Chai
Set in Hong Kong Park, LockCha serves Chinese teas and vegetarian dim sum in a dining room styled to resemble a scholar's quarters. There are free traditional music shows on Saturday evenings; reserve a seat. *lockcha.com; KS Lo Gallery, 10 Cotton Tree Dr, Wan Chai; dim sum from HK$28.*

ASIA

SINGAPORE

Singapore might be one of the world's most expensive cities, but that doesn't apply to the food. Once you've nailed the priciest part – finding somewhere to sleep – you can graze the city's food courts at prices you'll miss forever when you return home.

🖐 Hawker Chan Hong Kong Soya Sauce Chicken Rice & Noodle, Chinatown

A Michelin star thrust this hawker stall into the spotlight. The line forms before Mr Chan Hon Meng opens for business, and waiting times can reach two hours for the tender soya-sauce chicken and caramelised pork char siew. *02-126 Chinatown Complex, 335 Smith St; dishes S$2-4.*

🖐 Gluttons Bay, City Hall

This acclaimed row of al fresco hawker stalls is a great place to start your Singapore food odyssey. Get indecisive over classics like *laksa*, satay and barbecued stingray. Head in early to avoid a frustrating hunt for a table. *makansutra.com; 01-15 Esplanade Mall, 8 Raffles Ave; dishes from S$4.50.*

🖐 Satay Street, CBD

Each night, the street next to Telok Ayer Market is transformed into an open-air restaurant lined with sizzling barbecues. Order a dozen satay sticks doused in delicious peanut sauce and a jug of Tiger beer for the ultimate budget feast. *Boon Tat St; satay S$0.60 per stick.*

🖐 Bakery cafes, Tiong Bahru

After window-shopping, enjoy a pastry and a coffee from Tiong Bahru Bakery (*tiongbahrubakery.com; 107 & 252 North Bridge Rd*), Drips Bakery Café (*drips.com.*

sg; 82 Tiong Poh Rd) or Plain Vanilla Bakery (*plainvanillabakery.com; 1D Yong Siak St*)

🖐 328 Katong Laksa, Marine Parade

For a bargain foodie high, hit this cult-status corner shop. The star is the namesake *laksa*: rice noodles in a light curry broth made with coconut milk and coriander, and topped with shrimps and cockles. *328katonglaksa. sg; 51 East Coast Rd; laksa S$5.50-7.50.*

🖐 Timbre+, Queenstown

Welcome to a new era of hawker centres. With over 30 food outlets, Timbre+ has it all: artwork-covered shipping containers, food trucks, craft beer and live music nightly. But the food draws the crowds. *timbreplus.sg; 73A Ayer Rajah Cres; dishes from S$3.*

Below: colourful cafes and food stalls abound in Singapore.

ASIA

PENANG, MALAYSIA

Cultures and cuisines collide in Malaysia's most famous eating enclave. For the price of a latte back home, you can feast on everything from Chinese fried noodles and Indian tandoori chicken to sublime roti canai (flatbread with curry dipping sauce).

✋ Anjung Gurney Night Market

Penang's most famous hawker complex sits just past the Gurney Plaza mall. It throngs with stalls serving flame-cooked satay sticks, *loh bak* (five-spice pork sausage), *lok-lok* (skewered treats cooked in broth), Penang *laksa* (sour fish soup) and more. *Persiaran Gurney; mains RM5-15.*

✋ Restoran Kapitan

Perhaps the best tandoori chicken in town – with strong ginger and garlic tones and plenty of chilli heat – is served at this busy 24-hour canteen. Eat by the street in the fun, frantic open-fronted downstairs dining hall. *Lorong Chulia; dishes RM8-19.*

✋ Lorong Baru Hawker Stalls

Ask locals for their favourite hawker stalls and most will pick this night-time street extravaganza. It's a great spot for *ikan bakar* (grilled seafood), *asam* and Penang *laksa*, fried oysters, and *popiah* (spring rolls). *New Lane Pasar Malam, cnr Jln Macalister & Lorong Baru; mains RM3-20.*

✋ Sri Ananda Bahwan

This buzzy Indian *nasi kandar* (rice and curry) restaurant whips up penny-priced masala *dosas* (rice-and-lentil pancakes), *idli* (steamed rice cakes) and *roti canai* (flatbread with dips) for breakfast, plus more substantial rice plates and *thalis* (set meals).

facebook.com/srianandabahwan; 53-55 Lorong Penang; mains RM4-15.

✋ Pulau Tikus Hawker Centre

This lane off Jalan Burma is crammed every morning with locals filling up for the day on *char koay teow* (fried noodles with seafood), *nasi lemak* (pandan-leaf-wrapped coconut rice) and the like. A taxi from central George Town will cost RM15-20. *Cnr Solok Moulmein & Jln Burma; dishes RM5-20.*

✋ Penang International Food Festival

All the city's best restaurants get in on the act, with food stalls taking over whole streets of the old city. Feast on everything from curries to satay sticks and *ikan bakar*. *piff.com.my; Apr; mains from RM5.*

Below: Penang's culinary culture is a mix of Chinese and Malaysian influences – like its architecture.

ASIA

MUMBAI, INDIA

Mumbai is India's biggest, busiest and most cosmopolitan metropolis – and probably the priciest place to visit in all of India, too. However, there are plenty of ways to save, including when sampling the city's famous street food for spice without the price.

Bademiya Seekh Kebab Stall

Every evening, drivers cruise to this back lane in Colaba to feast on a delicious meat-heavy menu, eaten hot on the streets, or picnic-style on car bonnets under yellow street lights. Top treats include spicy, fresh-grilled kebabs and *tikka* rolls (kebabs in rolled bread) hot off the grill. *bademiya.com; Tulloch Rd, Colaba; meals from Rs 130.*

Hotel Ram Ashraya

In the Tamil enclave of King's Circle, just outside Matunga Rd train station, 80-year-old Ram Ashraya is beloved by southern families for its spectacular *dosas, idli* (spongy, round, fermented rice cake) and *uttapa* (pancakes with spicy toppings). *Bhandarkar Rd, King's Circle, Matunga; meals from Rs 40.*

Badshah Cold Drink

Opposite Crawford Market, Badshah has been serving snacks, fruit juices and its famous *falooda* (a rose-flavoured drink of milk, cream, nuts and vermicelli) to hungry bargain hunters since 1905. A must. *52/156 Umrigar Bldg, Lokmanya Tilak Marg, Lohar Chawl; snacks from Rs 38.*

Pancham Puriwala

Located just outside Chhatrapati Shivaji Maharaj Terminus, this budget eatery is a heritage icon, serving *puri bhaji* (puffed-up bread with a potato-and-pea curry) for over a century. Diners have been thronging here since before the first train services started in India. *8/10 Perin Nariman St, Ballard Estate; meals from Rs 40.*

K Rustom

Inside this modest looking hole-in-the-wall joint are just a few metal freezers, but the ice-cream sandwiches – served between two crisp wafers – have been pleasing Mumbaikar palates since 1953. Pick from some 50 flavours; roasted almond crunch is usually the bestseller. *87 Stadium House, Veer Nariman Rd, Churchgate; ice cream from Rs 40.*

Bhajia by the beach

Head to either Juhu Beach (north of the city) or Girgaum Chowpatty (right in the centre) at sunset, and join the crowds of promenaders munching on paper plates of *bhelpuri* (puffed rice tossed with fried dough, lentils, onions, herbs and chutneys) and *pav bhajia* (spiced veg with a buttered bun) by the sand. *Snacks from Rs 20.*

ASIA

CHIANG MAI, THAILAND

The northern Thai capital attracts gourmands by the busload, with its legendary night markets, lip-smacking street food and cooking courses. Feast on fine Northern Thai cuisine by night, then learn to make it the next day in a local farmhouse kitchen.

Khao Soi Lam Duan Fah Ham

North of Ratanakosin Bridge, Khao Soi Lam Duan Fah Ham is the best place to sample Chiang Mai's legendary *kôw soy* (crispy and soft wheat-and-egg noodles in curry broth). Crowds gather daily for delicious, filling bowls. *352/22 Th Charoenrat/Th Faham; mains from 40B.*

Huen Phen

Set in a bric-a-brac filled house that feels more like a garden, Huen Phen serves full-flavoured northern Thai food, from *nam prik noom* (vegetables dipped in incendiary sauces) to Burmese-style pork curry. Come for lunch, when queues are smaller and prices lower. *112 Th Ratchamankha; mains from 30B.*

Laap Kao Cham Chaa

Popularised by the late, great Anthony Bourdain, this low-key street stall beneath a rain tree is a staple for authentic northern Thai food. Plop down in a plastic chair, grab a Singha and feast on heaping plates of pounded-pomelo salad, Thai-style beef or pork *lâhp* (salad). *Th Ratanakosin; closed Sun; mains from 50B.*

Lert Ros

As you enter, you'll pass the main course: delicious whole tilapia fish, grilled on coals and served with sticky rice and a fiery Isan-style dipping sauce. The menu also includes fermented pork grilled in banana leaves and spicy green papaya salad. *facebook.com/LertRosRestaurant1986; Soi 1, Th Ratchadamnoen; mains from 30B.*

Huen Jai Yong

Devotees of Northern Thai food shouldn't miss this out-of-town gem, where authentic, hard-to-find dishes beloved by Lanna people are served in traditional ways. Try the tilapia or raw-pork *lâhp*, and a spicy soup with either (whole) frog or sun-dried fish and ant eggs. *64 Mu 4, Th San Kamphaeng; closed Mon; mains from 30B.*

Asia Scenic Thai Cooking

On Khun Gayray's economically priced Thai cooking courses you can study in Chiang Mai or at a peaceful out-of-town farm. Classes cover soups, curries, stir-fries, salads and desserts, so you'll be able to make a three-course meal after a single day. *asiascenic.com; 31 Soi 5, Th Ratchadamnoen; cooking courses 800-1000B.*

ASIA

TOKYO, JAPAN

Tokyo gets a bad rap for being expensive, but it isn't completely justified. Many of the top sights are actually free, and even feasting on the city's famous noodles and sushi won't empty the treasury if you know where the bargains lie.

🖐 Katsu Midori Sushi, Shibuya-ku

A spin-off from Tokyo sushi shop Sushi-no-Midori (sushinomidori.co.jp), this is the city's best *kaiten-zushi* (conveyor-belt sushi) restaurant. It's a bargain for the quality and it's always crowded, so you know the plates are fresh. *Seibu Department Store, 21-1 Udagawa-chō, Shibuya-ku; plates ¥100-500.*

🖐 Onigiri Yadoroku, Taitō-ku

Onigiri, rice moulded into triangles and wrapped in sheets of *nori* (seaweed), is Japan's ultimate snack. Try them made-to-order at Tokyo's oldest *onigiri* shop (opened in 1954). *onigiriyadoroku.com; 3-9-10 Asakusa, Taitō-ku; onigiri ¥310-760*

🖐 Kagawa Ippuku, Chiyoda-ku

This humble restaurant specialising in *Sanuki-udon*, wheat noodles from Kagawa in Shikoku, is a bargain. Pay at the vending machine; you'll be handed an English menu to help with the options. *kagawa-ippuku. com; Tokyo Royal Plaza,1-18-11 Uchikanda, Chiyoda-ku; udon ¥430-820.*

🖐 Forest Beer Garden, Shinjuku-ku

Many Tokyo rooftops host beer gardens in the summer – at Forest, there's a two-hour all-you-can-eat barbecue buffet coupled with all-you-can-drink beer. *14-13 Kasumigaoka-machi, Shinjuku-ku; May-Sep; dinner course women/men ¥3900/4200.*

🖐 Isetan Department Store, Shinjuku-ku

The food hall in this department store has outlets from some of the country's top restaurants. Create a meal of sushi, dumplings, *tonkatsu* (fried pork cutlet) sandwiches and cake then take it upstairs to eat on the roof garden. *isetan.mistore.jp/ shinjuku.html; 3-14-1 Shinjuku, Shinjuku-ku; dishes from ¥500.*

🖐 Sagatani, Shibuya-ku

This noodle joint possibly wins the prize for Tokyo's best cheap meal: fresh, stone-ground *soba* (buckwheat) noodles are made daily and served with a side of *goma* (sesame) dipping sauce. Wash it down with a ¥150 beer. *2-25-7 Dōgenzaka, Shibuya-ku; 24hr; noodles from ¥280.*

Below: Shinjuku-ku offers a choice of drinking dens with a smorgasbord of snack options to keep you going.

ASIA

FESTIVAL & EVENTS

Asia's festivals are not just events, they are rites of passage. Alongside big-ticket extravaganzas organised by tourist boards and city councils, a dozen religions have imprinted their diverse beliefs on the landscape, imbuing the festival calendar with more drama than a Bollywood musical.

PRACTICALITIES

The biggest hassle at festival time is finding somewhere to stay. Whatever the special occasion, rates double or triple and budget rooms book out months

Above: each colour at Holi represents something: red for love and fertility and green for new beginnings. Right: Hong Kong's Clockenflap festival.

ahead. For some showstoppers, finding a room at all can be a major challenge. It's not unheard of for people to stay hundreds of kilometres away, commuting in on the day by bus or train.

• **Plan your transport** – bus, train and plane seats can't be found for love nor money during major religious festivals such as Thailand's Songkran or India's Durga Puja. This can be a time when hiring a scooter or car pays dividends.

• **Unexpected shutdowns** – during Vietnam's Tet and Ramadan in Muslim regions, many businesses close and even finding a meal can be a challenge. Nothing operates at all during Bali's Nyepi (Day of Silence).

• **Look for pilgrim services** – at festival time, extra trains, planes, buses and boats are laid on for devotees, often a cheap route to the festivities if you don't mind the crush.

• **Be safe** – when big crowds gather, so do pickpockets and bag-grabbers; keep your belongings and hotel room secure.

We don't want to sound like your dad, but intoxicated travellers are particularly easy pickings. Women should be careful of wandering hands in crowds.

• **Protect your gear** – many of Asia's biggest festivals feature oceans of flying water; put any tech you want to keep safe in a waterproof bag or back in your hotel room. During Holi in India, everyone gets doused with coloured powder, so wear some clothes you don't mind getting ruined.

RELIGIOUS FESTIVALS

There's rarely a fee to attend Asia's big religious festivals, but millions of locals have the same idea, meaning transport can be rammed. For the biggest events, you might have to make travel plans six months or even a year in advance.

• **Research the area around the festival** – do you have to stay in the same town? Would it be cheaper to stay nearby and commute in by bus, train or rented scooter?

• **Know your calendars** – most Muslim, Hindu, Buddhist, Jain, Sikh and animist festivals are linked to the lunar calendar, so dates change every year.

• **Be a pilgrim** – pilgrimage is a key element of many celebrations, and joining the crowds is an intense experience; many festivals lay on special tent accommodation and cheap meals along the pilgrimage trails.

Religious festivals are often riots of colour, but never forget that these are deeply spiritual events for believers. Follow these tips to avoid upsetting participants.

• **Dress appropriately** – taboos on naked flesh go double at festival time: cover arms, legs and hair, if appropriate. Follow the lead of locals and you won't go far wrong.

TIBETAN BUDDHIST NEW YEAR

Forget fireworks; Buddhist monasteries across the Himalaya celebrate the Tibetan New Year – Losar – with spectacular dances, featuring squads of monks in vividly colourful costumes. Festivities spill over into the two-week prayer festival of Monlam, marked by dances in monstrous masks of protector deities, animals and celestial beings. Dances are free to watch at monasteries from Ladakh to Kathmandu and Sichuan to Bhutan. Dates depend on local traditions and the lunar calendar, falling sometime between January and March – check ahead when visiting Tibetan Buddhist regions.

ASIA

Above: after their moment in the sun, Harbin's ice sculptures melt back to their liquid state.

• **Festival etiquette** – travel 101 really, but some rituals and religious sites are reserved for believers, so seek local advice before joining the procession – and don't be intrusive with your camera.

MUSIC & CULTURE

Culture in Asia takes myriad forms, from free book readings at the Jaipur Literature Festival (jaipurliteraturefestival. org; Jaipur, India; January-February) to (expensive) ice-sculpture viewing at the Harbin International Ice and Snow Festival (Harbin, China; January-February; ¥135-330).

Asia's music scene is as diverse and fun-filled as the continent itself. K-Pop, hip hop, death metal, gamelan – whatever your genre, you'll find a festival celebrating it somewhere. Unlike back home, sleeping in a tent is not the norm – most events take place in designated venues and people go back to homes and hotels (typically fairly sober) at the end of the day.

Book decades in advance to secure tickets for decibel-busting parties such as Clockenflap (clockenflap.com; Hong Kong; November), Fuji Rock Festival (en. fujirockfestival.com; Japan; August) and Sunburn (en-gb.facebook.com/ SunburnFestival; Pune; December).

• **Look out for free concerts** – religious events often feature free traditional music, adding to the spiritual mood. Listen out for *qawwalis* (devotional hymns) on important dates at Sufi shrines across South Asia.

• **Seek out sponsored street parties** – promotion is big business in Asia, and free music events and pageants backed by drink and food companies are easy to find, particularly in Malaysia, the Philippines and Thailand.

• **Civic pride** – events organised by city,

• **Know the rules** – alcohol is often restricted, and some festivals ban eating, meat and even kissing. Marijuana use is tolerated for some Hindu religious festivals, but don't assume foreigners are automatically allowed to dive in.

state or national governments offer the best chance of free concerts and performances – traditional dance, music and theatre feature prominently.

Above: Songkran in Thailand is noted for the eager participation of locals and visitors.

FOOD FESTIVALS

Municipal authorities across the region pull out all the stops for showstopping public feasts of street food, and even famous restaurants tout their wares at food stalls down on the street. The best bit? Many are free to attend; you won't even pay elevated prices for the food.

Top picks include the Penang International Food Festival (piff.com. my; April), Singapore Food Festival (visitsingapore.com; June-July) and Hong Kong Food Festival (food-expo.com.hk; December).

SPORTS

The big events on the sporting calendar are celebrated with almost religious devotion. Getting a ticket for a cricket test match or international football tournament requires dedication and funds. There are normally some free-to-watch events at the Southeast Asian Games and other regional multi-sport competitions.

HIGHLIGHTS

BUN BANG FAI, VIENTIANE, LAOS

A riot of rockets fills the skies around Vientiane for this explosive rain festival, which welcomes, nay demands, the arrival of the monsoon rains. Some of the homemade missiles are taller than the people firing them! *p77*

HOLI, DELHI, INDIA

Wear clothes you can throw away for India's vivid festival of colour. Any and everyone is sprinkled in rainbow-coloured powder, which repaints the streets in a supersaturated spectrum of colour. Grab some powder and get hurling. *p78*

JAIPUR LITERATURE FESTIVAL, INDIA

This intellectual celebration of the spoken word is a great chance to see heavyweights from the literary world reading their own work, alongside debates from some serious world thinkers. After a few days of higher thought, you'll see the Pink City in a whole new light. *p79*

SONGKRAN, CHIANG MAI, THAILAND

The Thai New Year is a water-fight to beat all others. Water balloons, hoses, squirt guns...all are unleashed on observers and participants to mark the arrival of the Thai New Year, particularly around Chiang Mai's medieval moat. You won't stay dry, so why not join in the soggy, free fun? *p79*

ASIA

FESTIVALS

🖐 Baybeats, Singapore

A free music festival is a rare thing, particularly when it's one of Asia's biggest celebrations of indie music and alternative rock. Held at Esplanade – Theatres on the Bay, this three-day extravaganza pulls in talent from across Asia and further afield. *esplanade.com; Colonial District; Aug; free.*

🖐 Beijing International Kite Festival, China

Cast your eyes to the skies for this annual two-day celebration of a centuries-old Chinese pastime. Incredibly colourful and frequently complex creations are floated into the big blue above the capital. *Běijīng Garden Expo, Fengtai District and Olympic Green, Tongzhou District; Apr; free.*

🖐 Bun Bang Fai, Vientiane, Laos

All around the Lao capital, this spirited rain festival sees thousands of rockets fired into the air to celebrate and encourage the arrival of the monsoon rains. Some of the rockets can reach 9m in length and climb several kilometres into the sky. It's also celebrated vigorously in the Thai town of Yasothon. *Vientiane; May/Jun; free,*

🖐 Chingay, Singapore

On the 22nd day after Chinese New Year, Chingay delivers Singapore's biggest and best street parade – a flamboyant multicultural affair featuring floats, lion dancers and other cultural performers. Tickets are needed for official viewing galleries, but you can watch for nothing from the roadside barriers. *chingay.gov.sg; CBD; Feb; public spectators free.*

Left: launching lanterns from Pingxi in Taiwan. **Below:** flamboyant festival costumes at Chingay, Singapore.

ASIA

🖐 Diwali, India

Diwali (Deepavali) fills the night with light all over India. This Hindu spectacular symbolises the victory of the forces of good (light) over the forces of evil (darkness); accordingly, butter-lamps light up homes and megatons of fireworks burst into the night sky. *India-wide; Oct/ Nov; free.*

🖐 Fire Dragon Dance at Mid-Autumn Festival, Hong Kong, China

Over three nights during this festival, a 67m-long straw dragon decorated with 70,000 glowing incense sticks is paraded by hundreds of enthusiastic young men in the affluent Causeway Bay neighbourhood of Tai Hang. A rival dragon parades in Pok Fu Lam village. *taihangfiredragon.hk; Sep/Oct; free.*

🖐 Ganesh Chaturthi, Mumbai, India

There's no charge to watch Mumbai's biggest annual bash, when thousands of vividly coloured statues of Ganesh, the elephant-headed god of wisdom, are ritually cast into rivers, tanks (temple pools) and the sea – some are so hefty they have to be dunked by crane. *Aug-Sep; free.*

🖐 Hanami, Tokyo, Japan

When the cherry blossom blooms in early spring, the whole of Tokyo goes into party mode. Popular parks such as Yoyogi, Ueno and Inokashira fill with revellers and picnics. Entry to the parks is free; beer and snacks from the nearest convenience store are optional. *Late Mar-early Apr; free.*

🖐 Holi, Delhi, India

India's most colourful festival is a vivid, technicolour spectacle. Celebrating

Above: a statue of Ganesh is manoeuvred through Mumbai's crowds.

the victory of the forces of good over the demon Holika, devotees hurl *gulal* (coloured powder) at anyone within reach; wear some clothes you don't mind spoiling and join in the fun. *India-wide; Feb/Mar; free.*

Hungry Ghost Festival, Hong Kong, China

To appease roaming spirits when the gates of hell are unbolted, people make food and faux-money offerings by the roadside, while communities set up bamboo scaffolds to host Buddhist and Taoist ceremonies, and Chinese opera performances. *Aug/Sep; free.*

Indra Jatra, Kathmandu, Nepal

To honour Indra, god of rain, the Kumari (a young girl worshipped as a living goddess) is paraded through the streets during this famously frenetic festival marking the end of the monsoon. For a freebie, join the devotees sipping rice beer from the mouth of the Sweta Bhairava statue in Durbar Square. *Aug/Sep; free.*

Jaipur Literature Festival

This massive celebration of the written word is the largest free literary festival in the world; advance registration gets you access to readings (by such esteemed writers as William Dalrymple and Vikram Seth), debates and more. *jaipurliteraturefestival.org; Jaipur, India; Jan-Feb; free.*

Singapore International Film Festival

Singapore's largest film fair showcases both groundbreaking international films and home-grown cinema. Screenings (some free) are held at multiple venues across the island, including lots of flicks by new talent. *sgiff.com; various locations; Nov-Dec; some free screenings.*

Songkran, Chiang Mai, Thailand

The annual three-day celebrations for the Thai New Year have morphed into the world's biggest water fight, with high-powered super-soakers and water balloons being launched at any and all onlookers. The most intense water battles take place around Chiang Mai's medieval moat. *Apr; free.*

Taiwan Lantern Festival

Thousands of sky lanterns fill the sky over Taiwan's Pingxi district like swarming jellyfish during this annual extravaganza. There's no charge to watch the launching, or be deafened by the bursts of firecrackers at Yanshuei's Wumiao Temple. *Feb-Mar; free.*

Thaipusam, Kuala Lumpur, Malaysia

Kuala Lumpur's Batu Caves are the setting for one of Asia's most surreal spectacles, when thousands of Hindu pilgrims perform gruesome acts of self-mortification to honour Murugan, god of war, and parade to the holy caves from downtown KL. *Jan/Feb; free.*

Below: thousands of Tokyo citizens welcome the cherry blossom every March and April.

ASIA

OUTDOORS & ADVENTURE

In Asia, nature is a free playground, with everything from coral reefs and deserts to wildlife-filled jungles and raging rivers calling out to lovers of the great outdoors. There are even rainforests in the middle of cities, and cities in the middle of deserts.

An astonishing range of natural wonders can be witnessed for free, or at prices you won't mind paying when you're standing just feet from a wild elephant, orangutan, Komodo dragon or royal Bengal tiger. The thing to keep in mind is that every penny you do spend

Above: Railay in Thailand is a world-renowned rock-climbing destination. Right: floating in the Dead Sea is less demanding.

helps make nature worth more alive than dead, so you're investing in the future, not just in the moment.

NATIONAL PARKS & NATURE RESERVES

First things first – completely free national parks are like crouching tigers and hidden dragons: you won't find them. All the famous Asian national parks you've probably heard of have entry fees, and often those entry fees are higher for tourists than for locals. Every country has its own parks and rules, usually set by departments of wildlife or forestry.

The good news is that park fees are often modest compared to the thrills that await inside. Many smaller, less famous national parks and reserves charge refreshingly affordable fees, often pegged to a local's budget, and a few precious jewels have no entry fees at all, though guide fees may still apply. Don't overlook state-run forest parks,

which often have just as much nature at half the cost.

Entry fees are only part of the picture. Many national parks insist that visitors hire a guide, which usually means shelling out to hire a jeep, boat or another wildlife-proof means of transport. Riding in on elephant-back is rightly falling out of favour; more progressive parks will let you walk with elephants instead, a much more humane arrangement.

For pocket-friendly wildlife encounters, consider the following:

• **Add up the costs** – when comparing parks, look at all the costs, including entry fees, accommodation, guide fees, camera and video permits, park transport, and getting to the park in the first place.

• **Which safari?** – most parks can be visited multiple ways: by jeep, by boat, by raft, by canopy walkway, or even on foot. A free walk on a trail by the park HQ may not provide the same wildlife-spotting opportunities as a safari deep inside the reserve.

• **Tourists, assemble** – to avoid paying more as a solo traveller, gather your own group together to share the costs: ask around at places where you eat and sleep.

• **Put in the time** – a one-day, zip-in, zip-out tour may look great on your budget sheet, but is that enough to get close to the wildlife? It can take several safaris to get that once-in-a-lifetime encounter.

• **Look to the fringes** – many parks are fringed by buffer zones and migration corridors where visitor fees are lower, providing alternative territory for wildlife encounters.

• **Bring a dry bag** – they call it rainforest for a reason, and replacing a damp camera or phone can easily tip your trip into the red.

CAMPING IN ASIA'S NATIONAL PARKS

Accommodation in Asia's national parks can be lavish beyond your wildest dreams – or rudimentary, cheap and more atmospheric than you could ever have hoped for. Thailand's national parks (dnp.go.th) are particularly well set up for independent campers, with government-run sites at most national park headquarters charging from 30B per night (with your own tent), often with a cheap canteen for camp meals. Top picks include pitching a tent on the snow-white beaches in Ko Karutao Marine National Park and sleeping out in elephant country in Khao Yai National Park.

ASIA

ADVENTURE ACTIVITIES

With landscapes seemingly made for the purpose, Asia has made big business out of thrills and spills. Whole atolls are set aside as marine parks for diving and snorkelling, whole coasts are surfing and rock-climbing playgrounds.

However, be ready for back-home costs. US dollar prices are the norm, and you may not get much change from a US$100 bill for a full day of organised adventure. Seek out centres run by and for locals and you'll often save, as well as supporting small, independent businesses.

The first rule for cost watchers is bring your own gear, whether that means climbing ropes for Krabi, hiking boots for trekking in Nepal, or surfboards for Bali. Airlines are well used to adventurers checking in with surfboards, mountain-bikes and the like – just check the baggage rules ahead of time.

• **See underwater** – reefs can be accessed directly from thousands of beaches in Thailand, Malaysia, Indonesia and the Philippines; carry your own mask and snorkel and observe marine life for free.

• **Surf spectaculars** – OK, a surfboard is not the easiest thing to lug around, and it's easy to find boards for rent, but having your own means free access to hundreds of the world's top lefts and rights, particularly in Indonesia and Sri Lanka.

• **Take a tent** – trekking costs more if you need accommodation; a tent means you won't have to shell out for lodges or an organised trek. In Nepal, it will cut your trekking costs to just meals, the national park fees and trekking card.

• **Solo trekking?** – actually, common sense dictates trekking with at least one other person, but you don't always need to join an organised trip. Dozens of routes in Nepal have inexpensive teahouses

HIGHLIGHTS

RIDE THE MAHAKAM RIVER, INDONESIA
Public ferries offer impromptu rainforest safaris on the Mahakam River, which cuts into the heart of Borneo for 980km. Spot wildlife on the banks, stop off in remote townships and stay overnight in rustic homestays; it's a ready-made Borneo adventure. *p87*

HIT THE REEFS OFF GILI AIR
Indonesia's Gili Islands are the whole budget paradise package, and best of all, the reefs just offshore from those lovely beaches are in easy snorkelling range. Bring your own gear or rent locally and dive into a marine wonderland right off the sand. *p92*

CLIMB KRABI, THAILAND
The karst outcrops at Railay are a natural jungle gym, providing vertical kilometres of bolted climbing routes on limestone buttresses, cave-roofs and tufas. With no fees to climb, all you need to do is turn up with a harness, boots and rope and start climbing. *p90*

DESERT WONDERS AT LITTLE PETRA, JORDAN
Timeless ruins under the stars in a desert moonscape...and it's free? Jordan's Siq Al Barid, aka Little Petra, is everything you'd want from the desert, minus the crowds and cost. When night falls, sleep in Bedouin camps under an amazing canopy of stars. *p91*

dotted along the trails, offering rooms for US$5 or less, and many trails are easy to follow without a guide.

• **Stay where you play** – sports such as kite-surfing and scuba diving are centred on residential camps where tuition often goes hand-in-hand with cheap rooms and meals.

• **Getting to adventures** – a big part of the cost of organised adventures is transport: travel to the cliffs, or dive centre, or rafting put-in point independently and rates can plummet.

• **Tour by public transport** – often, a public bus, train or boat provides the same vantage point as an organised tour for a fraction of the cost.

• **Guide or no guide?** – taking a guide supports the local economy, but visiting national parks and wildernesses you can explore independently will cut your costs (stay in homestays so you're still contributing).

Top left: an Indian elephant in Kaziranga National Park. Bottom left: boat tours in the rainforest of Taman Negara, Malaysia. Above: walking in Wadi Rum, Jordan.

ASIA

ASIA'S
BEST
DISCOUNT WILDLIFE ENCOUNTERS

Nature in Asia is wild and untamed, with tigers – and monkeys, and elephants, and rhinos – burning brightly in the jungle night. Entry fees are nominal compared to the wonders that await.

01

03

01

KAZIRANGA NATIONAL PARK, INDIA

Assam's Kaziranga offers pretty much guaranteed sightings of one-horned Indian rhinos, tramping through the elephant grass around marshy pools. The park is home to two-thirds of Asia's surviving rhinos, which easily justifies the entry and jeep fees. *kaziranga. assam.gov.in; Assam, India; entry Rs 650*

02

ULU TEMBURONG NATIONAL PARK, BRUNEI

Spectacular Ulu Temburong National Park is best accessed from the charmingly low-key Sumbiling Eco Village, partly run by Iban tribal people. Guests can often get discounts for helping out at the lodge, and rates include accommodation, jungle treks, riverboat rides – and blowpipe darts! *sumbiling.com; Kampong Sumbiling Lama, Jln Batang Duri, Brunei; US$140*

03

TAMAN NEGARA NATIONAL PARK, MALAYSIA

Malaysia's Taman Negara National Park has wildlife in droves (including elephants, tigers and leopards), and inexpensive buses run from KL to Jerantut, where you can connect with park transport. Explore cheaply on foot and cross the legendary canopy walk, 45m above the forest floor, for RM5. *tamannegara.asia; Taman Negara National Park, Jerantut; park entry RM30.*

04

ORANGUTAN ENCOUNTERS IN BORNEO

For low-cost encounters with the old man of the forest, head to Sepilok Orangutan Rehabilitation Centre in Sabah (Jln Sepilok, Sandakan; admission RM30), or head to Sakau Village and hire a boat and driver for a two-hour cruise (RM100) on the Kinabatangan River to spot orangutans and other uniquely Bornean critters in the wild.

05

KOMODO DRAGONS AT RINCA, INDONESIA

Seeing living dinosaurs at Komodo National Park is the stuff of dreams, but the park fees mount up fast. Wise travellers head to nearby Rinca Island, where admission includes three guided walks to spot Komodo dragons (and their prey) in their natural habitat. *Rinca Island, Indonesia; admission 80,000Rp*

06

KHAO YAI NATIONAL PARK, THAILAND

A wealth of options for independent travellers push Khao Yai to the top of the wallet-friendly list. Sure, there's an entry fee, but many park trails are open to hikers without a guide, you can rent bikes for road exploring, plus there's a night market for discount eats, and cheap camping and hostel accommodation nearby. *khaoyainationalpark.com; Khao Yai National Park, Thailand; entry 400B.*

07

SEEING TIGERS IN KANHA NATIONAL PARK, INDIA

Kanha Tiger Reserve is one of the best places on the planet to spot tigers, and you can save money by travelling to Khatia village by local bus and getting together with other visitors to share a jeep and guide. *Khatia, Madhya Pradesh, India; entry per jeep passenger Rs 260, guide fee Rs 360, jeep rental Rs 2375.*

BORNEO

Asia's largest island is a wild and untamed garden, draped in rainforests, traversed by jungle rivers, and teeming with wildlife found nowhere else on earth. But gratifyingly, getting up close to unspoiled nature doesn't have to cost the earth.

⚉ Explore the Danum Valley, Malaysia

The field centre in the verdant, wildlife-filled Danum Valley Conservation Area welcomes nature enthusiasts as well as researchers, with moderately priced hostel accommodation and camping, and endless kilometres of steamy jungle trails where there are chances of spotting orangutans and jungle cats. Rangers will show you the trails for RM30 per hour. *danumvalley.info; Danum Valley Conservation Area, Sabah, Malaysia; dorm beds/camping RM95/80.*

⚉ See the 'Blue Tears' at Miri, Malaysia

There's no charge for one of Borneo's most otherworldly spectacles – the 'blue tears' of bioluminescence at Tusan Beach, just half an hour from Miri. Created by tiny algae called dinoflagellates, the neon glow that lights up the night from September to December could be plucked straight from Avatar. The only cost is getting here. *Tusan Beach Subis, Sarawak, Malaysia; free.*

⚉ Climb into cathedral-sized caves at Niah, Malaysia

Unlike the famous caves at Gunung Mulu, the limestone caverns at Niah National Park can be reached by bus and taxi, cutting down overheads. There's no need to take a guide to explore the towering caverns with their swooping swiftlets and bats, who gush in and out of the caves at dusk, and there's cheap hostel and camping accommodation at the park HQ. *Niah National Park, Sarawak, Malaysia; entry RM20.*

⚉ Take the bus to Bako, Malaysia

Unlike most national parks in Borneo, Bako National Park near Kuching is accessible by public bus (for just RM5). Disembark in a lush, green wonderland, where well-marked jungle trails offer the chance to spot everything from pitcher plants to proboscis monkeys. There's no charge to trek, but the boat ride to the park HQ costs RM40. *sarawakforestry.com; Bako National Park, Sarawak, Malaysia; park fee RM20.*

Below: you might find that seeing an orangutan in Malaysia inspires you to support these threatened primates.

Spotting proboscis monkeys in Brunei

You don't need to leave Bandar Seri Begawan, Brunei's capital, to get close to wildlife. Walk to the city limits and scan the trees – or even better, charter a river taxi from the waterfront to go a little way upriver – and you'll see Borneo's second most famous primate swinging through the jungle trees. Boatmen charge B$30-40 for an hour cruise, best taken at first light or around sunset. *Bandar Seri Begawan, Brunei.*

Soaking in the Poring Hot Springs, Malaysia

There's no cheap way to climb Mount Kinabalu, but afterwards (for a nominal fee) you can soak tired muscles in the Poring Hot Springs, developed by the Japanese in WWII. A dozen or so pools and tubs are fed with sulphurous water from a steamy underground spring. *sabahtourism.com/destination/kinabalu-national-park; Kinabalu National Park; Sabah, Malaysia; entry RM15.*

Flop on Miri's sands, Malaysia

Beaches trace the coast of Malaysian Borneo like a gold garland, and getting to a pristine strip of sand is easy almost anywhere along the coast. Rent a car or motorcycle in Miri and explore the side-roads running down to the shore from the coastal Q642 highway heading south towards Bintulu. *Miri, Sarawak, Malaysia; free.*

Public ferry cruises on the Mahakam River, Indonesia

Forget Malaysia-style organised boat trips; from the gateway city of Samarinda in Indonesia's Kalimantan, you'll pay less than US$25 for the 37-hour journey by public ferry along the wide Mahakam River to Long Barun. En route, you can scan the jungle for wildlife and leap off at tribal townships, stopping overnight in simple homestays. *Samarinda, Kalimantan, Indonesia; fare to Samarinda 350,000Rp.*

Snorkel off Semporna, Malaysia

Divers spend fortunes reaching the legendary dive sites of Sipadan, but you can snorkel the nearby reefs of Tun Sakaran Marine Park from Semporna – and see a similarly wonderful array of marine life – for a fraction of the cost. Arrange trips to Pulau Mataking, Pulau Mabul and Pulau Kapalai islands with Uncle Chang's (ucsipadan.com) or Big John Scuba (bigjohnscuba.com). *Semporna, Malaysia; snorkelling trips from RM200.*

Budget trips up the Kinabatangan River, Malaysia

Tracking for 560km through dense rainforests, this mud-brown river serves as Borneo's most evocative transport route, offering sightings of monkeys, orangutans, hornbills, crocodiles and herds of pygmy elephants. Stay in community-run homestays such as Miso Walai Homestay (mescot.org; rooms from RM70) and arrange trips with local boat owners to see the same wildlife wonders for less. *Batu Putih, Sabah, Malaysia.*

Beach bumming near Kota Kinabalu, Malaysia

The Malaysian town of Kota Kinabalu has easy access to some of Borneo's finest strips of sand. Tanjung Aru, by the airport, is a favourite destination, but the water is cleaner and the setting lovelier on the paradisiacal islands of Tunku Abdul Rahman Marine Park. To get here, jump the ferry from Jesselton Point Ferry Terminal in KK. *Sabah, Malaysia; park entry RM20.*

BORNEO TRAVEL TIPS

Borneo is split between Malaysia, Indonesia and (slightly more expensive) Brunei. Save on transport by taking slow but cheap buses and boats over internal flights, and by using Grab and other rideshare services or renting a motorcycle for shorter bespoke trips. Many national parks have campgrounds, hostels and homestays, and local boat trips are often cheaper if negotiated directly with boat owners. Eat where locals eat (and what they eat) and your overheads will drop even further.

ASIA

BEST WET & WILD ACTIVITIES ON A BUDGET

You don't need a private yacht to appreciate the deep, blue sea in Asia: even budget travellers can get in on the diving, surfing, snorkelling, kayaking and rafting action.

SURFING AT ARUGAM BAY, SRI LANKA

For those just getting steady on their surfing feet, Sri Lanka's laid-back Arugam Bay is the ideal place to get board-confident with cheap rental and lessons, abundant cheap accommo-dation, cheap eats and bathtub-warm waters. *Arugam Bay; board rental Rs 1000, lessons from Rs 2500.*

ECONOMY DIVES AT PULAU PERHENTIAN, MALAYSIA

With a dive certif-icate in hand, the world is your oyster. Peaceful Pulau Perhentian has long been the budget dive destination of choice, with boat and beach dives from RM90, and snorkelling trips from just RM40. Stay cheaply in beach 'chalet' resorts, with dorm beds in stilt huts. *Pulau Perhen-tian, Terengganu.*

LEARNING TO SCUBA DIVE ON KO TAO, THAILAND

Thousands of would-be divers head to Southeast Asia every year to get certified, and lovely Ko Tao is one of the cheapest spots on earth to do it, with plenty of competition and teeming reefs close offshore. A four-day PADI Open Water course costs as little as B10,000, and tast-er fun-dive courses start from B2500. *Ko Tao, Thailand*

WHALE SHARK ENCOUNTERS AT DONSOL, PHILIPPINES

At Donsol, huge numbers of *butand-ing* (whale sharks) migrate by from November to June, and snorkellers can get astonishingly near to these gentle giants – an amazingly close encounter for the cost. *Donsol, Bicol; shark spotting trips per six-person boat P3800.*

SNORKELLING OFF THE GILI ISLANDS, INDONESIA

Diving is never cheap but snorkelling can take you into the same magical world for the cost of hiring a mask, snorkel and fins. You'd struggle to find a lovelier spot than the idyllic Gili Islands, with cheap backpacker rooms and rainbow reefs starting just 100m offshore. *Gili Islands, Lombok; mask and snorkel hire 50,000Rp per day.*

SEA KAYAKING AROUND HALONG BAY, VIETNAM

The karst outcrops of Halong Bay and nearby Lan Ha Bay off Cat Ba island were once the haunt of pirates; get closer to that romantic vision by skipping the or-ganised cruises and paddling yourself around instead. Sea kayak rental starts from US$35 for the day, opening up se-cret bays, limestone grottoes and stilt villages.

KITE-SURFING ON KALPITIYA LAGOON, SRI LANKA

Dozens of castaway-style kite-surfing resorts around millpond-flat Kalpitiya Lagoon can train you up, rent you gear, let you loose on the lagoon, and furnish you with a bed under a whirling ceiling fan at the end of a wet and wild day. Try laid-back Kite-surfing Lanka (www.kitesurfinglanka.com; Kandakuliya, Kalpitiya Peninsula).

RAFTING THE TRISULI RIVER, NEPAL

The rivers that thunder down from the Himalaya provide some of the world's best – and cheapest – whitewater rafting, with numerous put-in points within easy reach of Kathman-du. The gushing Trisuli gushes just below the Kathman-du-Pokhara road, reducing costs for a day's rafting to US$40 per person.

© TonyNg | Shutterstock

THAI PENINSULA

The narrow isthmus of land stretching south towards Malaysia is Thailand's playground, from the island-studded bays around Phuket to the sea cliffs of Krabi and the underwater wonderland off Ko Tao. Rope up, paddle on or dive in.

Karst adventures at Ao Phang-Nga

Ao Phang-Nga – secret hideout of the Man with the Golden Gun – is a maze of caves and karsts. Rent a stand-up paddleboard for self-guided exploration from Paddle Phuket (paddlephuket.com). *Ao Phang-Nga, Phuket; paddleboard rental from 500B.*

Climbing at Krabi

Kilometres-worth of beachside limestone climbing routes at Railay and Ton Sai have been sport-bolted, opening access to big walls, overhangs and cave roofs. If you need to rent gear, try King Climbers (railay. com; Walking St, Railay). *Free.*

Snorkel Ko Tao

Watery Ko Tao is a famously cheap place to scuba dive, but there's also great snorkelling, accessible just offshore for the cost of renting a mask and snorkel. Spot small sharks and other impressive fauna at Laem Thian, Ao Leuk, Ao Tanot and Shark Island. *Gear rental 100-200B per day.*

Explore Chiaw Lan Lake by boat

The dense rainforests of Khao Sok National Park hide everything from bears and boars to elephants and rarely spotted tigers, meaning maximum wildlife for your buck. Guided walks (1200B) are one way to explore, or charter a boat (2000B per day) to investigate the sawtooth shore of the park's Chiaw Lan Lake. *khaosok.com; Khao Sok National Park.*

Full moon parties at Ko Pha-Ngan

Everyone knows when and where the party is on Ko Pha-Ngan: the full moon, at Hat Rin. Up to 40,000 partygoers gather for this hedonistic all-nighter, and while it's no longer free, you'll struggle to find a bigger party for a lower price. Entry 100B.

Find your own version of *The Beach*

The inspiration for Alex Garland's novel is off-limits but it's possible to find similar shores all over Thailand, for the cost of a chartered longtail boat. Idyllic Ko Tarutao Marine National Park is a good place to start; camp at Ao Son. *Satun; park fee 200B.*

Below: the Full Moon Party at Hat Rin beach may be slightly more regulated these days but it's a still a blast.

© 4FR | Getty Images

ASIA

SOUTHERN JORDAN

An island of calm in a turbulent region, easygoing Jordan puts the desert at your fingertips, with epic treks, wadi runs, lost cities, dune driving and more, all accessible to travellers on even the most modest of budgets.

✋ Petra without the price tag

Petra is worth every dinar, but there's no charge at all to enter nearby Siq Al Barid, aka Little Petra, where temples and ancient cave-homes stand whittled out from the rock. Bedouin camps offer inexpensive overnight stays. *Wadi Musa; free.*

✋ Float on the Dead Sea

Technically, it's free to swim (float?) in the Dead Sea. Most visitors access the waters at pricey resorts so there's somewhere to wash off the salt afterwards; there's only a moderate charge to access the beach, pools and showers at the Amman Beach resort. *Dead Sea Hwy; entry JD25.*

✋ Wadi Rum on the cheap

The JD70 Jordan Pass gets you a free visa (normally JD40) and free entry to 40 sites including wonderful Wadi Rum. Even without the pass, the tiny entry fee is a small price to pay to hike through this alien desertscape, where rocky outcrops emerge like sandworm fangs from the desert floor. *Wadi Rum, Aqaba; entry JD5.*

✋ Dark sky watching

Venturing into the desert almost anywhere in Jordan delivers perfect conditions for star-gazing, but some sites are particularly spectacular: the small entry fee to Dana Biosphere Reserve delivers views deep into

the Milky Way. *rscn.org.jo/content/dana-biosphere-reserve-1; Dana; entry JD8.*

✋ Snorkelling at Aqaba

Beach hotels at Aqaba offer access to Red Sea reefs, often for an eye-watering price. A tip: bring your own mask and snorkel and take a public bus (JD0.50) to South Beach, south of Aqaba, where the reef rises yards offshore. At Seven Sisters, there's even a submerged tank teeming with fish. *Free.*

✋ Cycle the King's Highway

This back road runs north through the desert towards Madaba. Tackling it by bike, you'll be immersed in the landscape; carry plenty of water and recharge at the Crusader castles at Karak and Shobak. *Free.*

Below: Wadi Rum sparks the imagination like few other places and can be enjoyed with only a modest fee.

ASIA

BALI, LOMBOK & THE GILI ISLANDS, INDONESIA

Generations of surfers, divers and seekers of sun, sea and sand have flocked to Bali, Lombok and the Gilis for a satisfyingly economical diet of noodles, grilled lobster, all-day surfing and diving, island culture and late-night beers on the beach.

Surf Bali's budget breaks
Kuta has the fame, but there are quieter breaks all along Bali's wave-buffeted shoreline. Rent a board locally and hire a scooter (from US$5), allowing easy access to everywhere – from the accessible surf at Canggu to stuff-of-legends Padang Padang on the Bukit Peninsula. *Board hire from US$5; lessons from US$30.*

Freedive the Gili Islands
Scuba diving is old hat. The only gear you need to freedive is a mask, snorkel, fins and two generous lungfuls of air. For those new to the sport, courses on Gili Trawangan cost a similar amount to learning to scuba dive – from US$290 with Freedive Gili (freedivegili.com) – but once skilled-up, you can dive deep into the underwater world anywhere without the need for tanks.

Climb Mount Rinjani, Lombok
The ascent of Mount Bromo may get all the thunder, but Lombok's Mount Rinjani is 1400m taller, and is also a sacred site. Climb to the crater lake in two challenging days from Sembalun – a cheaper trip than the longer hikes up most Indonesian volcanoes. *rinjaninationalpark.com; guided treks US$75-100 per day.*

Dive and snorkel off Gili Air
The Gili Islands get another nod thanks to the easy snorkelling off Gili Air, whose whole east coast is brocaded by reefs just 100m to 200m offshore. Guesthouses and dive shops rent out snorkelling gear at bargain prices, and you can wade in off the sand anywhere along the shore. *Gear hire 50,000Rp per day.*

Kick back on sublime sands
The smallest Gili island, Gili Meno, has the best of the chain's beaches, particularly at its southeast corner. For more stunning strips of sand, head to western Lombok: Mangsit is a laid-back, budget surf hub; or try delightfully undeveloped Tanjung Aan or Mawun Beach on the south coast. *Free.*

Below: Bali's waves range from beginner-friendly breaks to pro-level barrels off the Bukit Peninsula.

ASIA

© Wonderful Nature | Shutterstock

CHIANG MAI PROVINCE, THAILAND

Temples, fab food, amazing outdoor activities – is there anything Chiang Mai doesn't have? The northern Thai capital is the gateway to a mountain province with jungle-cloaked peaks, hill-tribe villages and a lifetime's worth of low-cost outdoor fun.

Trek to Doi Suthep

The trek up Doi Suthep mountain – crowned by Chiang Mai's most famous monastery – starts near the TV tower behind the city zoo. Climb forested slopes to the calm, meditative Wat Pha Lat monastery and continue along the river to sacred Wat Phra That Doi Suthep. It's a sweaty hour-long walk with a big reward at the end. *Free.*

Mountain-bike down a mountain

The rainforest-draped mountains around Chiang Mai offer thrilling descents for those willing to put in the effort to cycle up first (or you can pay agencies to drive you to the top). Rent a mountain bike from Chiang Mai Mountain Biking (mountainbikethailand.com) and pedal up to Wat Phra That Doi Suthep, then hit the forest tracks. *Bike hire per day 250B.*

Climb Crazy Horse Buttress

Run by passionate climbers, Chiang Mai Rock Climbing Adventures (thailandclimbing.com) were responsible for setting up most of the bolted routes on the towering Crazy Horse Buttress. With your own gear, you can take advantage of their daily shuttle bus to the outcrop (395B return) and climb for free to your heart's content.

Motorcycle through the jungle

You don't have to roam far from the city to reach the jungle. Branching west at Mae Rim, Rte 1096 climbs into the forested Mae Sa Valley. The road follows a winding loop past waterfalls and fruit farms flanked by forest, a rewarding 100km round-trip from Chiang Mai by rented motorcycle. *Motorcycle hire from 150B per day.*

Climb a waterfall

Nam Tok Bua Tong (Sticky Waterfall) earned its name for the calcium deposits left by the mineral-rich waters. Scrambling up the grippy boulders with water gushing past is a blast, and you can continue uphill to the forest spring that provides the miracle minerals. *65km north of Chiang Mai; free.*

Below: the forests of Chiang Mai are filled with waterfalls, mountain-bike and hiking trails and suspension bridges.

ASIA

© fredfroese | Getty Images

ASIA'S
BEST
LOW-COST TREKS

Part of the thrill of exploring Asia's most remote corners is getting there on foot. Trek independently to snowpeaks, volcanoes and jungles and keep trekking costs to a minimum.

ANNAPURNA CIRCUIT TREK, NEPAL

Local buses buzz to the trailheads for Nepal's other big-ticket trek (after Everest Base Camp). With a TIMS card and Annapurna Conservation Area Project permit (NRs 3000), independent trekkers can bus from Pokhara to Besisahar or Bhulbhule and just start walking, emerging two to three weeks later at Jomsom having looped right around the mighty, snow-capped Annapurna Massif.

EVEREST BASE CAMP, NEPAL

There are cheap teahouses all along the two-week walking route to Everest Base Camp, so you can hike without joining an organised trek. A Trekkers Information Management System card (which lets you trek everywhere in Nepal) costs NRs 2000, national park fees are NRs 3390, and overnight beds at teahouses cost from US$5.

THE GAUMUKH/TAPOVAN TREK, INDIA

The big lure of this famous pilgrimage route in Uttarakhand, northern India, is not the stunning scenery but the spiritual significance. The two-day route climbs from Gangotri to the source of the River Ganges at Gaumukh, and on to high-altitude meadows at Tapovan. *Permits, camping and cooking fees around Rs 1000, porters cost from Rs 800.*

MARKHA VALLEY, LADAKH, INDIA

The beauty of this breathless, high-altitude trek through an arid Himalayan mountainscape is the lack of prep required. Trekkers can board a local bus to Chilling and start walking, staying overnight in local homestays that charge Rs1000 for dinner, a bed, breakfast and a packed lunch for the next day's walk.

VOLCANO TREKKING IN THE PHILIPPINES

Dozens of the Philippines' many volcanoes can be climbed (after first checking no eruptions are imminent) by asking guesthouses to put you in touch with local guides. Skip the touristy trek to Pinatubo and dangerous Taal Lake in favour of Hibok-Hibok on Camiguin Island, and Mt Pulag and Mt Bulusan on Luzon. *Guides P1200 per day.*

TRIBAL ENCOUNTERS AT BAC HA, VIETNAM

Swap commercialised Sapa for peaceful Bac Ha, whose weekend market attracts traders from 11 different tribal communities from surrounding villages in the terraced highlands. Staying in local homestays, you can walk at your own pace to villages or arrange guides locally (from US$55 per day) for longer day- and overnight trips.

TREK THE GREAT WALL, CHINA

Once you get beyond the beautified sections near Beijing, the unrestored Great Wall stretches endlessly to the horizon. A good taster trek is the 20km walk from the Jiankou section, near Xizhazi village, where the only cost is a RM20 fee to enter the village itself (you can wild camp among the ruined and toppled battlements).

© Fat Jackey | Shutterstock

WELLNESS

Asia may be the birthplace of yoga, meditation and massage, but locals are not giving away their wisdom for free. Many places tread a fine line between spiritual centre, hospital and commercial resort. Almost all expect participants to pay fees for therapies – a free lunch can be found; finding a free massage is harder.

Save cash by creating your own cost-free wellness experiences – waterfall swimming, sunset watching, dawn tai chi, yoga on the beach – and spend money on Ayurveda (Indian traditional medicine), massage and other things you can't do for yourself.

Above: practising yoga in public in Bangkok's Suan Luang Rama IX park. Right: meditation (here in Hampi, India) doesn't cost a cent.

YOGA, MEDITATION AND SPIRITUALITY

Yoga is found wherever travellers gather, from the beaches of Indonesia to the high Himalaya. India, Nepal and the Buddhist nations of Southeast Asia are top spots for travellers seeking wellness with a spiritual angle.

• **Visiting any Buddhist monastery or temple** is an opportunity to participate in meditation; respectful visitors are welcome to sit in on prayer sessions across the Himalaya and Buddhist Southeast Asia.

• **Wellness at source** – sacred towns such as Rishikesh and Bodhgaya in India are major centres for yoga and meditation, with numerous meditation centres and ashrams that welcome international practitioners for (often) moderate fees.

• **Find free classes** – yoga and meditation centres usually charge a fee, but many hotels and guesthouses offer free sessions for guests. Look out for cheap classes advertised on noticeboards in traveller hangouts.

• **Hospital or spa?** – spas aimed at foreign travellers always cost more, sometimes vastly more; seek out centres that define themselves as hospitals or clinics, aimed at local practitioners.

MASSAGE AND AYURVEDA

Wat Pho in Bangkok was the birthplace of Thai massage. Appropriately, Buddhist monasteries are some of the best places to find inexpensive, properly executed Thai massage, with the added bonus of knowing there'll be no funny business. And don't overlook Japan's famous onsens and other hot springs for a more passive form of relaxation.

In India, Ayurveda is the big story. This traditional form of medicine uses everything from a vegetarian diet to massage, herbal remedies and enemas (OK, not for everyone) to effect a whole-body cure. Prices at Ayurvedic spas and hospitals can be sky-high; look for clinics unaffiliated with resorts unless you want to pay luxe spa prices.

• Guesthouses, hotels and resorts can hook you up with a reputable masseur at a fair price. In the Philippines, many blind people are trained as masseurs, offering their services at airports, bus stations and ferry terminals.

• In Thailand's Chiang Mai, former prison inmates retrain as masseurs to provide a legitimate career after release at centres such as Lila Thai Massage (chiangmaithaimassage.com).

• Tibetan Buddhism has its own form of traditional medicine, with inexpensive consultations and treatments available at the Men Tsee Khang (men-tsee-khang.org) hospitals in Dharamsala and Leh in India.

• Ayurveda is a luxury affair in Sri Lanka; Colombo's Siddhalepa Ayurveda (siddhalepa.com) and Unawatuna's Sanctuary Spa (sanctuaryspaunawatuna. com) offer treatments at more affordable prices.

• Chinese traditional medicine is highly accessible, with pharmacies offering walk-in consultations and herbal treatments wherever Chinese communities are found.

HIGHLIGHTS

BATHE IN 'SIN DEMOLISHING' WATERS, INDIA

Keralan beach resort Varkala doubles as a sacred site thanks to its mineral-rich springs emptying into the ocean. Swimming here is said to remove sins, so splashing about on the beach is a path to both physical and spiritual well-ness. *p101*

SWEAT IT OUT IN A LAO HERBAL SAUNA, LAOS

There's a moderate charge for a tradi-tional herbal sauna at Luang Prabang's Lao Red Cross – a small price to pay for a therapy with this kind of wow. The combi-nation of steam and medicinal herbs will leave you giddy with exhilaration. Enjoy a herbal tea or a massage afterwards. *p102*

FREE THINKING

01

FREE YOGA BY THE GANGES, RISHIKESH

Attending a formal yoga ashram in Rishikesh typically means strict lifestyle conditions: no smoking, alcohol, meat or sex. For more casual practitioners, Sivananda Ashram has free drop-in classes daily, taking place at 6am on the banks of the River Ganges when Rishikesh is at its most serene. *sivanandaonline. org; Lakshman Jhula Rd; free*

02

POYA DAYS, SRI LANKA

When thronged by white-robed Buddhist pilgrims on *poya* (full-moon) days, every temple, shrine and dagoba in Sri Lanka hums with spirituality. Great free spots to observe *poya* days include the imposing Kelaniya Raja Maha Vihara (Kelaniya; free), just outside Colombo, and ancient Ridi Vihara (rideeviharaya. lk; near Kurunegala; free).

03

TAI CHI, SHANGHAI, CHINA

Shanghai's grand, colonial-era Bund is one of the most popular settings for locals to practice tai chi chuan, as the sun climbs above the Huangpu River and its flanking forest of skyscrapers. It's a fascinating combination of the ancient and ultra-modern that tells you a little about what motivates China today. *Zhongshan East 1st Rd; free*

04

MEDITATION IN BODHGAYA, INDIA

Where better to practise meditation than the place where the Bhudda attained enlightenment? Bodhgaya is crammed with monasteries and meditation centres; to get involved, drop into the Root Institute for Wisdom Culture (rootinstitute. ngo; Saxena Rd) for free dawn meditation sessions at 6.45am.

05

It seems almost incongruous to pay for spirituality – luckily, you can experience almost every Asian therapy, philosophy and spiritual activity for free (or for small donations).

06

07

05

THAI MASSAGE, WAT PHO, THAILAND

To find a real, non-dodgy Thai massage, head to its spiritual source. Wat Pho (Th Sanam Chai, Bangkok) is both an astonishing temple and a Thai massage school, where you can get a relaxing 30-minute pummelling from monastery-trained practitioners, all for a wallet-pleasing 260B.

SAND MANDALA MAKING, THIKSEY GOMPA, INDIA

Thiksey's timeless gompa (Buddhist monastery) is one of the most spectacular settings in which to view the creation of sand mandalas – intricate drawings made from coloured sand which take days to create and are then destroyed as a reminder of the impermanence of existence. There's a tiny fee to enter the 15th-century monastery. Thiksey, Ladakh; R30

MONK CHAT, CHIANG MAI, THAILAND

The myriad monasteries of Chiang Mai are fascinating to explore at any time, but during the daily Monk Chat sessions, you can sit down and discuss *dharma* with the resident novices. Drop by Wat Suan Dok (Th Suthep; free, entry to main hall 20B) on Monday, Wednesday and Friday afternoons.

08

AEROBICS AT SARANROM PARK, BANGKOK

After a few seafood feasts, many visitors to Bangkok feel like they might benefit from a little exercise. Why delay? Every evening, Saranrom Park fills with the sound of grunts and pop beats as free open-air aerobics classes roll into action.

GOA, KARNATAKA & KERALA, INDIA

India has a strong claim to being the birthplace of wellness – at the very least, this was where yoga and meditation were first conceived – and the southwest states of Goa, Karnataka and Kerala offer rich pickings for travellers seeking healing on a budget.

✋ Sunrise Yoga, Goa

If you already know the postures, a dawn yoga session on the beach is positively de rigueur in Goa. If you need some coaxing to bend into more challenging poses, keep in mind that many guesthouses offer free classes for guests, and centres such as Swan Yoga Retreat (swan-yoga-goa.com) in Assagao accept drop-in practitioners for Rs 350-500.

✋ Receive a temple blessing, Kerala

You don't have to be a believer to feel a frisson when receiving a blessing at a centuries-old Hindu temple such as Thiruvananthapuram's Shri Padmanabhaswamy (Fort, Thiruvananthapuram, Kerala; free). A donation is usually expected, and the ceremony ends with a waft of flame, a sprinkling of water and a sacred mark applied to the forehead with coloured *kumkum* power.

✋ Have a hug, Kerala

We're not kidding. Matha Amrithanandamayi Mission (amritapuri.org; Amrithapuri, Kerala) near Kollam is the seat of one of India's few female gurus, Amrithanandamayi, known as 'The Hugging Mother' because of the *darshan* (audience) she offers, where thousands of devotees get a comforting hug in marathon all-night sessions. *By donation.*

✋ Watch sunrise on a Goa beach

West-facing Goa is perfectly oriented to catch the sunset, but when it comes to watching sunrise, there's only one choice. Hollant Beach, near Bogmalo, sits at the mouth of a creek emerging between the palms; it's the perfect spot

Below: experience an Ayurvedic hot oil treatment in Kerala.

to feel the warm glow of the sun saluting the new day. *Bogmalo, Goa. Free.*

🤚 Ayurveda without the high price tag, Kerala

Kerala is a famous centre for Ayurveda – India's ancient system of herbal medicine – but prices at the state's most famous centres can raise the blood pressure, not lower it. For therapies without the sting, try the low-key Kannur Ayurvedic Centre (ayurvedawayanad. com; Kalpetta) where Ayurvedic massages start from just Rs 1000.

🤚 Wash away bad deeds, Kerala

Mineral springs are imbued with special spiritual significance in India. At Varkala, the healing waters emptying into the sea were revered as sacred long before tourists discovered the beaches. Bathing on golden Papanasham Beach is refreshing for body and soul – the mineral waters are said to wash away misdeeds, hence the name, which means 'demolish sin'. *Free.*

🤚 Meditate in your own personal temple, Karnataka

Inland from the Karnataka coast, the ruined temples of the Vijayanagar Empire tumble across an endless boulder-field at Hampi. Spirituality seekers with a head for heights can free-climb to ancient stone meditation shelters crowning rocky outcrops, providing maximum privacy and silence for contemplation. *Hampi, Karnataka; free entry to most areas.*

🤚 Stay Vegan, Goa

Even nonveg travellers will be blown away by the taste and variety of the food in Vagator's Bean Me Up cafe: vegan pizzas, homemade ice creams, tofu curry and dishes made from coconut, quinoa, tempeh, lentil dhal and cashew milk or cashew cheese. You can stay in rooms around the shady courtyard and join free morning yoga classes – the whole Goa package. *beanmeup.in; 1639/2 Deulvaddo, Vagator, Goa; meals from Rs 200.*

🤚 Swim in a waterfall, Kerala

Few things remind you of your place in nature like bathing in a waterfall, and Kerala is blessed with dozens of ideal cascades. At Soochipara Falls (aka Sentinel Rock) in Wayanad, white water surges over three tiers, thundering down into a deep plunge-pool in the forest that's perfect for swimming – it's the ultimate power shower. *Near Kalpetta, Wayanad, Kerala; entry Rs 50.*

🤚 Taste the food of the gods, Kerala

Prasad, sacred food offered to the gods, is doled out to devotees at temples across Kerala, offering a little taste of the divine. At the 13th-century Ambalappuzha Sree Krishna Temple near Alappuzha, pilgrims queue daily to offer their devotions and taste the temple's legendary *paal payasam*, a delicious milky porridge prepared using rice, milk and sugar. *Ambalappuzha, Kerala; free entry.*

🤚 360° wellness in Gokarna, Karnataka

The pilgrim town of Gokarna covers all the bases. You can stay inexpensively near the sand and try Ayurvedic treatments such as *shirodhara* (warm oil drizzled on the forehead) at Arya Ayurvedic Panchakarma Centre (ayurvedainindien.com; Kudle Beach; rooms from Rs 1200), take yoga classes at Shree Hari Yoga (shreehariyoga. in; Kudle Beach; classes Rs 300), surf the breaks or just relax on the sand.

EAT YOUR WAY HEALTHY FOR LESS

South Indian cuisine is famous for its lavish use of spices, and many of those spices have surprising medicinal effects. Cinnamon is a famous stomach settler, but it also helps lower blood sugar, reducing the risk of diabetes. Cloves have proven anti-inflammatory and antibacterial properties (extract of cloves is widely used in modern dentistry), turmeric contains powerful antioxidants, and cumin is a rich dietary source of iron, potassium and zinc. Something to think about while feasting on South India's bargain-priced vegetarian *thalis* (plate meals with multiple sauces).

ASIA

LUANG PRABANG REGION, LAOS

Laid-back Luang Prabang is where Laos comes to unwind, whether meditating in monasteries, swimming in waterholes, observing the sunset or swinging in a hammock watching the river flow by. Abundant cheap digs and affordable food seal the deal.

☝ Meditate in a monastery

Famous monasteries such as Wat Mai Suwannaphumaham and Wat Xieng Thong have entry fees, but dozens of smaller ones are free to all. Head to Wat Sensoukaram (Th Sakkarin) or Wat Xieng Mouane (Th Xotikhoumman) early in the morning and join the novices for morning prayers. *Free.*

☝ Watch sunset from Phu Si Hill

Prepare your legs for a steep ascent to reach this hilltop above town, crowned by a gilded stupa. There are epic 360° views across the Mekong and Nam Khan rivers, and Luang Prabang's houses, monasteries and temples. *Entry 20,000K.*

☝ Watch the river from a hammock

You'll feel your mental health improve with each gentle swing of your hammock as you watch the swirling waters of the Nam Khan flow by from under a palm thatched canopy at Fan Dee Guesthouse. It's the quintessential Lao way to relax. *Ban Vieng Mai; dorm/room from US$5/25.*

☝ Take a herbal sauna

The venerable Lao Red Cross is the place to come for traditional wellness. As well as first-rate massages, there's a terrific sauna infused with medicinal plants. *Lao Red Cross, Th Wisunarat; sauna 15,000K, massage from 50,000K.*

☝ Splash in the pools of Tat Kuang Si

There's a small fee to visit Tat Kuang Si, 30km southwest of Luang Prabang, but you get a lot of splashes for your kips. Gurgling waterfalls tumble over limestone formations into a series of cool, swimmable turquoise pools. It's a great spot to wash away both your worries and the sticky heat. *Tat Kuang Si; entry 20,000K.*

☝ Vegetarian grazing at a night market

Lao food offers rich pickings for vegetarians. Head to the night market behind the tourist office, where stalls sell healthy, wholesome meat-free dishes such as coconut pancakes, curried jackfruit and *láhp* (Lao salad) with tofu. *Th Sisavangvong; dishes from 15,000K.*

Below: the waterfalls of Tat Kuang Si, south of Luang Prabang, are very popular; arrive early for a cooling dip.

© Kim Briers | 500px

ASIA

BALI, INDONESIA

Indonesia's favourite aquatic playground, beautiful Bali is also a stunning spot for healing therapies, though a massage can come with a stressful price tag. Seek out smaller spas and local experiences to enjoy Bali wellness on a budget.

Yoga for free

Some Bali yoga centres feel like luxury resorts and charge similar prices, but there are still cheap classes around. A free early morning beachside class takes place regularly on Sanur Beach near Denpasar (near the end of Pantai Karang), or there's a more formal morning session at Lapangan Puputan Renon park in Denpasar town. *Free.*

Affordable pummelling

Ubud is crowded with spas offering an astonishing range of relaxing massage therapies, but prices may make you wince. For a more painless massage, try Nur Salon (nursalonubud.com) set in a traditional Balinese compound filled with labelled medicinal plants. *Jl Hanoman 28, Ubud; massage from 175,000Rp.*

Experience silence

During the annual Hindu festival of Nyepi in March, all of Bali falls silent. Twenty-four full hours are set aside for self-reflection, and everything stops, from music to markets, from traffic on the island's roads to flights into the airport. Nobody works or plays, so use the time as locals do for silent contemplation. *Free.*

Spring-water swimming at Air Sanih

The freshwater springs of Air Sanih are best visited at sunset, when throngs of locals bathe under blooming frangipani trees by the shore. Believed to be both sacred and health-giving, the spring waters are channelled into a series of swimming pools before flowing into the sea; it's spirituality at its most utilitarian. *Near Singaraja; entry 20,000Rp.*

Enter the Gates of Heaven

More than 1700 steps link the seven temples that climb the side of 1058m Mt Lempuyang, but pilgrims make a beeline for Penataran Lempuyang to pass through the *candi bentar* (split temple gateway), marking the transition from the physical world to the spiritual one. Don't miss the views of sacred Mt Agung. *Mt Lempuyang, near Amlapura; donation 10,000Rp*

Below: at dusk the pools of Air Sanih in Singaraja are soothing and serene.

ASIA

THE BEST THINGS IN LIFE IN EUROPE: TOOLKIT

Europe can be a frighteningly expensive place to travel – or a bargain breeze, if you know the budget ropes. Take advantage of cut-price transport, free days at museums and galleries, and fee-free national parks and your Euros will take you far.

A trip to Europe can be as expensive as you make it. It's easy to clock up a heart-stopping bill if you eat at the top restaurants, sleep in the best hotels and take in all the top sights. The flip-side to this jetsetter's vision of Europe is a cornucopia of free museums and national parks, bargain train passes and cheap meals on every street corner.

Navigating this conglomeration of 44 nations (the 27 countries of the EU, plus the UK, Russia and assorted small states) means hopping between these two worlds, offsetting expensive stops with bargain days out, and taking full advantage of the astonishing variety of sights and experiences crammed into such a compact geographical area.

TOP BUDGET DESTINATIONS IN EUROPE

For every millionaire's playground there's a budget backwater where money goes miles, or rather kilometres – here's our cheap travel top five:
• **LONDON:** Free museums by the dozen, offsetting high living costs.
• **CZECHIA:** Bargains abound once you leave Prague, the costly capital.
• **SLOVENIA:** Alpine adventures without the elevated prices.
• **ALBANIA:** Europe's new frontier, blending East and West at discount prices.
• **TURKEY:** Great value for spectacular sights, beaches and food.

Below: with lots of cycle paths, Europe (here Verona in Italy) is a great place to explore by bicycle, whether you bring your own or rent a bike.

© Giordano Nicola | Shutterstock

TRANSPORT

Transport may not be your biggest travel cost in Europe, thanks to trains and buses that cross international borders like they were just lines on a piece of paper. However, prices vary significantly from country to country, and some journeys – a peak-time, same-day train ticket in the UK, for example – can induce heart palpitations.

Prices for long-distance transport climb steadily the closer you get to the day of departure, before shooting for the stratosphere on the actual day of travel. Travelling off-peak and seeking out discount cards for urban transport and passes for mutliple trips will stop travel costs spiralling out of control.

AIR

The footloose days of budget airline breaks every weekend have had a wake-up call from climate change, but flying is still an efficient way to hop from country to country – though many prefer to travel by land to see all the amazing landscapes.

• **Budget is best** – you may not get comfort or a big baggage allowance, but you do get bargain rates as low as €20 on some city-to-city hops. Easyjet (easyjet.com) and Ryanair (ryanair.com) rule the skies, but local budget carriers offer similar deals, so shop around.

• **Know where you're going** – many budget airlines fly to airports that are miles out of town, adding to your overall transport costs.

• **Get a pass** – all the big airline groups have Europe flight passes offering savings. Compare prices for OneWorld's Visit Europe pass (oneworld.com), SkyTeam's Go Europe pass (skyteam.com) and Star Alliance's Europe Airpass (staralliance.com).

TRAIN

A monument to joined-up thinking, the Interrail (www.interrail.eu) scheme covers train travel in 33 countries plus Eurostar (eurostar.com), meaning you can pick your way across the continent with a single pass. You need to be resident in Europe to qualify; the Eurail Pass (eurrail.com) offers a similar deal for international visitors.

Interrail's Global Pass covers all 33 countries for a fixed number of days or months (priced from €185-677), but there are also Country Passes covering single nations. All offer savings on normal fares, and under-27s get extra discounts of up to 25%.

The value of either pass depends on how much you use it; for less frequent train travel, buying individual advance tickets (but never tickets on the day of travel) can still be cheaper. Student discounts can be generous; look out, too, for special railcards for young people and families offering bonus savings. Consult

Above: the Eurostar service from London to France, Belgium and the Netherlands and other high-speed rail services have transformed travel around the continent.

EUROPE

the Man in Seat 61 (seat61.com) for tips on buying tickets in every European nation.

BUS

International bus services zip all over Europe at bargain prices (even cheaper than trains), but journeys take longer and offer monotonous views of motorway traffic. The big players are Flixbus (flixbus.com), Eurolines (eurolines.de), BlaBlaBus (ouibus.com) and National Express (nationalexpress.com).

• **Budget bus deal** – Flixbus (flixbus. com) has a great 'Interflix' pass covering five long-distance trips in Europe for just €99, and their Megabus brand (megabus. com) has city-to-city fares from as little as €1.

CAR & MOTORCYCLE

Car-rental is cheap and easy across Europe, though the same can't always be said for petrol. Hire can cost less than €30 per day in many countries, making car rental cheaper per mile than the bus or the train on long journeys.

• **Pick the right class** – Europe's cheapest rental cars are tiny: sometimes barely with space for two adults and a suitcase. Go a class up to 'Economy' or 'Class B' and you'll get a cheap car you can actually fit into.

• **Cheap rental tips** – small local companies offer savings over international chains; avoid airport pick-ups and one-way rentals, or pay extra for the privilege. Book car hire with your inbound flight for discount rates.

• **Know the rules** – some rental companies allow you to drive into other countries, some don't. Note that some countries require a 'vignette' showing you've paid local road taxes, with fines if you drive without.

Above: camping, here in Lahemaa National Park in Estonia, can be a cost-effective way of passing the night in wild places – but check the rules for each country.

BOAT

Boats tie the UK and Ireland to Europe, join islands to the mainland, and link neighbouring nations along their coasts. Many car ferries (and Eurostar's UK-Europe car train service; eurotunnel.com) are priced per vehicle, with no extra charge per passenger, so travelling as a group can mean big savings.

LOCAL TRANSPORT

Taking conventional taxis will ramp up your costs; luckily, public transport (by buses and underground and overground trains) is so good in most European cities that you rarely need to say the 'T' word.
• **Carry the card** – most big European cities have pre-paid card systems covering all public transport, offering savings on the individual fare rate.
• **Look for tourist cards** – many cities offer sightseeing cards that include discounted or free urban travel, including Rome's Roma Pass (romapass.it), the Barcelona Card (barcelonacard.org) and Paris Pass (parispass.com).
• **Rent a bike** – many cities have cheap bike-share schemes, where you can pick up a bike and drop it off anywhere in the city; most trains carry bikes for free and cycle lanes make day-trips out of town a breeze.
• **Uber it** – Uber rules the roost in Europe, but MyTaxi, Lyft and BlaBlaCar are catching up fast.

ACCOMMODATION

Rates for hotels and guesthouses can be eye-watering, and in smaller places an expensive hotel may be the only option in town. Save by booking cheap rooms ahead of time, staying in hostels or home-shares from companies like Airbnb, or camping.

BUDGET SLEEPING

• **Camp out** – designated camping grounds in Europe usually charge, but free wild camping is permitted in parts of Sweden, Norway, Ireland, Estonia, Latvia and Spain (though not on private land). Elsewhere, look for economical campgrounds by the coast or at national parks.
• **Hostel a-go-go** – Europe has hundreds of chain and independent hostels, plus the slightly institutional hostels affiliated with Hostelling International (hihostels.com). Try Hostelworld (hostelworld.com) for continent-wide options.
• **Be unconventional** – find an even cheaper bed by Couchsurfing (couchsurfing.com), or joining a home exchange scheme such as Love Home Swap (lovehomeswap.com).

HOTELS, B&BS & GUESTHOUSES

• **Go local** – many countries have their own schemes for inexpensive rural accommodation: self-catering *gîtes* in France (gites-de-france.com), *agriturismo* in Italy (agriturismo.it), *hoevetoerisme* in the Netherlands and Belgium (hoevetoerisme.eu).
• **Go basic** – discount hotel chains are found all over Europe: rooms are simple, prices are minimal. Try Motel One (motel-one.com), Hotel Formule 1 (hotelf1.accor.com) and Ibis Budget (ibis.accor.com).
• **Be a B&B-er** – rates at B&Bs (*chambres d'hôte* in France) consistently undercut prices at hotels. More informal rooms and whole houses for rent are available on Airbnb (airbnb.com) and Homestay (homestay.com).
• **Rent a villa** – in the Med, private villa rental can be cheaper per day than resorts and hotels, and you'll build on the savings by self-catering.

PRO TIPS

I've rambled all over Europe on fun, family and football-related journeys as well as for work, and Europe doesn't have to be expensive. Some of the world's great treasure troves are either always free (London's British Museum) or free with certain conditions. EU nationals under 25 get a whole host of places including the Louvre in Paris without charge. Always carry ID and ask. Otherwise, just look at how locals live cheaply and follow suit. You're never far from an inexpensive bakery and travel is excellent value, especially if arranged in advance. Booked far enough ahead, a train from Brussels to Prague can cost as little as €29, for a journey of 866km!

Tom Hall, Lonely Planet

EUROPE

EUROPE'S BEST
TOURS BY PUBLIC TRANSPORT

Take a seat with the locals and embrace the trams, buses and boats that provide a cheap and interesting way to explore.

KUSTTRAM, BELGIUM

The 'Coast Tram' follows Belgium's North Sea shoreline for 67km, from De Panne to Knokke-Heist. Some 67 stops dot the route, enabling a bargain Belgian beach adventure. *Single €2.50, day pass €7.50.*

TRANSPORT CARD, GENEVA, SWITZERLAND

Pretty but pricey Geneva can make the impecunious weep. To ease the blow, the city will do you a deal: stay overnight and get a free Geneva Transport Card, covering travel on buses, trams and taxi-boats. *Free.*

TRAM 2, BUDAPEST, HUNGARY

Trundle alongside the Danube with views of the spires and turrets of Castle Hill, the Jewish Memorial (sculpted shoes on the riverbank) and grandiose parliament. *Single 350HUF, day pass 1650HUF.*

NUMBER 11 BUS, LONDON, ENGLAND

Ride a regular bus for a squeezed-in-with-the-natives view of the metropolis. Route 11 runs from Fulham to Liverpool Street Station via Chelsea, the Houses of Parliament, Trafalgar Square, St Paul's Cathedral and the Bank of England. *Single £1.50, day pass £4.50.*

BATEAU BUS, MONTE CARLO, MONACO

Monte Carlo's electric-powered ferry boats offer a smidgen of waterborne glitz in this razzmatazz harbour. They might not be billion-dollar yachts, but they offer the same views, and leave you with change to splash at the casino. *Single €2, day pass €5.50.*

TRAM 28, LISBON, PORTUGAL

Clatter around Lisbon's steep, tight-packed Alfama neighbourhood aboard a classic yellow tram. The Remodelado streetcars (designed in the 1930s) that work route 28 are the best way to get into the district's medieval alleys; the trams can get packed so go early. *Single €3, day pass €6.40.*

MERSEY FERRY, LIVERPOOL, ENGLAND

Hop aboard Europe's oldest ferry service on the 10-minute Seacombe-Pier Head commute, and as the Royal Liver Building's twin towers come closer, just try to stop yourself bursting into song... *Morning and evening Mon-Fri; single £2.80, return £3.70.*

BOSPHORUS FERRY, İSTANBUL, TURKEY

Travel between Europe and Asia for less than a Euro? Ferries crossing the Bosphorus Strait do just that, linking both sides of this continent-straddling city. Sail at sunset to see the minaret-pierced skyline in silhouette. *Single 4TL.*

VAPORETTO, VENICE, ITALY

You'll have to forgo the crooning gondolier, but opting for a public *vaporetto* (water taxi) instead of a private punt will save big bucks, while still offering views of Venice from water level. *Single €7.50, day pass €20.*

ARTS & CULTURE

Europe's big-ticket museums offer a
masterclass in human creativity, but
viewing these wonders can either cost an
arm and a leg, or not a penny, depending
on where you go. Plotting a route that
takes in Europe's free museums and
galleries is a prudent way to see the

sights without plunging into the red.

And seeing art doesn't just mean hours
in indoor museums. Many of Europe's
most striking cultural sights are open to
all for free: cities transformed into living
galleries by street artists; forlorn Soviet
statues on show in Moscow's Art Muzeon

**Above: the ruins at
Ephesus in Turkey.
Right: Castlerigg
stone circle in
Cumbria, England
is an alternative to
Stonehenge.**

Park; protests daubed on sections of Cold War-era walls in Berlin and Prague.

EUROPE'S TOP SWITCHEROOS

Seeing all of Europe's top flight museums, galleries and ruins would take a lifetime, and a hefty trust fund. Consider making the following trades.
- **SKIP:** Stonehenge, Wiltshire, England (entry €23.15)
- **SEE:** Castlerigg Stone Circle, Cumbria, England (free)
- **SKIP:** The Louvre, Paris, France (entry €15)
- **SEE:** Tate Britain, London, England (free)
- **SKIP:** Pompeii, Naples, Italy (entry €15)
- **SEE:** Ephesus, Selçuk, Turkey (entry €6.40)
- **SKIP:** St Paul's Cathedral, London, England (entry €22)
- **SEE:** St Peter's Basilica, Rome, Italy (free)
- **SKIP:** Château de Versailles, Versailles, France (entry €18-20)
- **SEE:** Peterhof, St Petersburg, Russia (entry €7.80)

EUROPE'S CULTURE FOR LESS

Free entry (or a day of free entry) is a surprisingly common feature at Europe's top museums and galleries, though donation boxes are prominently displayed to drop a hint. Historic sites are a different

story; there's normally a charge, and it can be as weighty as the ancient stones you're paying to see.

The positive take is that Europe has so many grand churches, venerable museums and ancient ruins that it's easy to find somewhere free, or almost free. In the process, you'll avoid the worst of the temper-testing crowds – meaning photos of the actual sights, not the tourists posing in front.

Look out for free days – even museums and galleries that aren't free every day are sometimes free one day a week (or at certain times every month); this includes the Hermitage in St Petersburg (free on the third Thursday every month), Madrid's Museo del Prado (free evenings) and other top bucket-list institutions.

Flash your ID – student savings are generous; bring your home student ID or ISIC card (isic.org). Many museums offer free entry for EU citizens aged under 25 (carry your passport or ID card).

Become a member – membership of a national heritage organisation, while costly, gives free entry to hundreds of historic sites, meaning it can quickly pay for itself. Start with the UK's National Trust (nationaltrust.org.uk) or English Heritage (english-heritage.org.uk).

Become a friend – for museums and galleries, 'friends' are members with benefits: for an annual fee, you get free access to paid-for temporary exhibitions and discounts in the shops and cafes.

Be a card-carrier – dozens of European cities have tourist-card schemes that offer free or discounted entry to the big sights (and discounted transport), including Amsterdam (iamsterdam. com), Rome (romapass.it), Barcelona (barcelonacard.org), Paris (parispass. com), Berlin (visitberlin.de) and Vienna (viennapass.com).

HIGHLIGHTS

NATURAL HISTORY MUSEUM, LONDON, ENGLAND

Every British kid's favourite museum, the Natural History Museum overflows with dinosaurs, stuffed beasties and shimmering crystals, proudly displayed in one of Europe's most eye-catching buildings (check out the fossils, fish and flora moulded into the walls). We defy you not to leave with at least a plastic diplodocus from the gift shop. p112

BASILICA DI SAN MARCO, VENICE, ITALY

The most famous landmark in the city on the lagoon, this famous basilica is, rather amazingly, free to visit, though the queues can be interminable. Inside this confection of domes and mosaics are wonderful Byzantine icons and artefacts, some plundered from Hagia Sophia during the Crusades. p116

EUROPE

LONDON

The Big Smoke may have cleaned up its environmental act, but it's not unusual to witness visitors choking on the prices charged for food, drink and beds. Smart travellers navigate around the tourist traps and concentrate on the magnificent free museums.

✋ British Museum

Who would want to miss the Rosetta Stone, key to deciphering Egyptian hieroglyphics (head upstairs for the mummies), the controversial Parthenon sculptures, the treasures from the Anglo-Saxon ship burial at Sutton Hoo and more? Founded as a 'cabinet of curiosities', this cultural epic is now quite simply the best history museum in the world. *britishmuseum.org; Great Russell St; free.*

✋ National Gallery

It started with three-dozen paintings in 1824 – now you can gawp at more than 2300 works from the world's finest brush- and chisel-wielding artists: Monet's *Water-lily Pond*; Van Gogh's *Sunflowers*; Rembrandt's portraits; Constable's renditions of pre-industrial England; and myriad masterpieces from da Vinci, Michelangelo, Raphael, Rubens, Turner and Cézanne. *nationalgallery.org.uk; Trafalgar Sq; free.*

✋ Natural History Museum

Every British child visits this building at least once, and so should you – it's worth it for the high-Victorian architecture alone, never mind the 80 million dinosaur skeletons, crystals, stuffed and mounted specimens and one of London's best museum shops. *nhm.ac.uk; Cromwell Rd; free.*

✋ The Science Museum

A cathedral to the technical achievements of humanity, with 300,000 objects spanning everything from aviation to microchips. Highlights include Stephenson's Rocket (one of the earliest steam trains), a full-size replica of the Eagle moon-lander from 1969, and the 'Clock of the Long Now', designed to keep time for 10,000 years. *sciencemuseum. org.uk; Exhibition Rd; free.*

✋ The Tate

This legendary gallery has two venues in London: Tate Britain (Millbank), displaying British art from 1500 to the present day; and fun-filled Tate Modern (Bankside) in

Below: Trafalgar Square and the National Gallery in the heart of London, the world's best city for free museums.

a converted power station, which shows wild, wacky and often controversial art from 1900 onwards. *tate.org.uk; free.*

The Wallace Collection
Artworks in this collection (including heirlooms snaffled from emergency sales as nobles fled the French Revolution) include Frans Hals' *The Laughing Cavalier* and works by Rembrandt, but it's as interesting to browse the weapons and armour and fine French furniture, including an oak commode that was once perched upon by Louis XV. *wallacecollection.org; Manchester Sq; free.*

Grant Museum of Zoology
Before private museums, many of the world's wonders were hidden away at universities. University College London's zoological museum is a window back to that time, with 68,000 specimens including a thylacine (Tasmanian tiger) and dodo. *ucl.ac.uk/culture/grant-museum-zoology; University St; closed Sun; free.*

Highgate Cemetery
London's most famous graveyard houses numerous dead celebrities (from Karl Marx and Malcolm McLaren to George Eliot and Douglas Adams), attracting many literary pilgrims. The west section is only accessible via guided tours (costly), while for the east cemetery there's a charge, but it's not too – ahem – stiff. *highgatecemetery.org; Swain's Lane; over-18s £4.*

Museum of London
Romp through 450,000 years of London history at this brilliant museum, one of the capital's highlights. Interactive displays and reconstructed scenes transport visitors from Roman Londinium and Saxon Lundenwic right up to the 21st-century metropolis. There's a partner museum in Docklands (West India Quay). *museumoflondon.org.uk; 150 London Wall (but relocating in next few years); free.*

The Wellcome Collection
Visit 'the free destination for the incurably curious' and discover how doctors treated people who'd tumbled into the Thames by blowing smoke up their bottoms. Established by 19th-century pharmacist Sir Henry Wellcome, this museum explores the junction of medicine, life and art. *wellcomecollection.org; Euston Rd; closed Mon; free.*

Victoria and Albert Museum (V&A)
Thanks to a collection of 4.5 million objects filling 145 galleries and spanning 5000 years, the V&A is the world's greatest art-and-design museum. It's an eclectic mix, from paintings, sculptures and photographs to textiles and crockery, covering every place and time from dynastic China to 1970s Britain. *vam.ac.uk; Cromwell Rd; free.*

Mediatheque
The British Film Institute's Southbank den is full of big- and little-screen gold. As well as (paid-for) movie screenings and lectures, don't miss Mediatheque, where you can simply park yourself in front of a widescreen computer and access thousands of documentaries, films and TV programmes. *bfi.org.uk; BFI Southbank, SE1; free.*

The Vault
Squirrelled away in a secure room (a former Coutts Bank vault), London's original Hard Rock Cafe keeps a hoard of musical memorabilia, including Kurt Cobain's sunglasses, a guitar strummed by Jimi Hendrix and a US Army shirt worn by John Lennon. Access is granted for nothing if you buy a burger and fries. *hardrock.com/cafes/london; 150 Old Park Lane; admission free.*

SURF THE SOUTHBANK

Flanking the Thames, the Southbank Centre is a vortex of cultural activity, both indoors and out. The Queen Elizabeth Hall Roof Garden is a summer oasis (and there's a bar); the Square hosts weekend street food markets; and the Riverside Terrace has live shows (in addition to the many buskers). Come summer, kids cavort in the jumping jets of Jeppe Hein's Appearing Rooms fountain, and at river level there's an urban beach. *southbankcentre. co.uk; admission free.*

EUROPE

PARIS

For the price of a drink, you can spend hours on a cafe terrasse watching the eternal theatre of Paris unfold. If you're able to drag yourself away, there are countless other free sights in this love-locked city of priceless glamour.

Cimetière du Père Lachaise

This hillside cemetery is known for celebrity graves, with Jim Morrison's, Oscar Wilde's and Édith Piaf's being the most visited. There are many others – writers Molière, Proust and Colette, and countless revolutionaries and politicians. It's a sculpture garden, historic monument and one of Paris' major sights. *pere-lachaise.com; 16 rue du Repos & blvd de Ménilmontant; free.*

Basilique du Sacré-Cœur

Begun in 1875 in the wake of the Franco-Prussian War, Sacré-Cœur is a symbol of the struggle between the conservative Catholic old guard and secular, republican radicals. It was finally consecrated in 1919, standing in contrast to the bohemian lifestyle that surrounded it. The views over Paris are breathtaking and, remarkably, there's only a charge to climb the dome (€7). *sacre-coeur-montmartre.com; Parvis du Sacré-Cœur; free.*

Deyrolle

This taxidermy shop dating from 1831 is a veritable museum of natural history. See animals exotic and familiar, great and small, incredibly lifelike and artistically posed (all from zoos, circuses or farms, naturally deceased). And all for sale, if you can fit a recumbent fox into your luggage (or budget). *deyrolle.com; 46 rue du Bac; closed Sun; admission free.*

Parc du Champ de Mars

Taking the lift to the tip of the Eiffel Tower can crunch the budget, but views from below are almost as stunning. Formerly a military parade ground, Parc du Champ de Mars offers manicured lawns that are one of Paris' best spots for a picnic; come with wine, cheese and a blanket and wait for the free show at dusk as the La Tour Eiffel lights up like a candle. *Champ de Mars, 7e; free.*

Musée Carnavalet

One of the best reasons to visit this

Below: lazing about outside Sacré-Cœur in Montmartre is free for all.

EUROPE

museum of Parisian history is its gorgeous setting in two 16th-century *hôtels particuliers*, lavish townhouses built by noblemen of Le Marais. Among its fascinating artefacts and artworks are the painstakingly reconstructed fin-de-siècle drawing rooms and Baroque interiors, glimpses into how the other half lived. *carnavalet.paris.fr; 23 rue de Sévigné; free.*

Mémorial de la Shoah

The haunting Memorial to the Unknown Jewish Martyr has metamorphosed over time into the Memorial of the Shoah. Museum exhibitions reveal the awful events that followed the German occupation of France during WWII. The actual memorial to the victims stands at the entrance, inscribed with the names of 76,000 men, women and children deported from France to Nazi extermination camps. *memorialdelashoah.org; 17 rue Geoffroy l'Asnier; closed Sat; free.*

Art 42

Street art and post-graffiti now have their own dedicated space at this 'anti-museum', with works by Banksy, Bom.K, Miss Van, Swoon and Invader (who's behind the Space Invader motifs on buildings all over Paris), among other boundary-pushing urban artists. Compulsory guided tours lead you through subterranean rooms sheltering some 150 works; reserve a place in advance. *art42.fr; 96 blvd Bessières; tours every 2nd Tue; free.*

Musée d'Art Moderne de Paris

Perhaps the best of Paris' less-known museums, with an outstanding permanent collection of 20th-century artworks, Art Deco furniture and more. Standouts are Raoul Dufy's *La Fée*

Electricité (a room-sized mural depicting the discovery of electricity), Matisse's *La Danse* and a collection of Robert Delaunay's masterpieces. *mam.paris.fr; 11 ave du Président Wilson; closed Mon; free.*

Musée de la Vie Romantique

Walk through the paved courtyard of this 1830 *hôtel particulier* (grand townhouse) and into the artistic life of 19th-century Paris. Chopin, Delacroix, Ingres, Liszt and others frequented Ary Scheffer's Friday-night salons here, and the museum features Scheffer's works and the memorabilia of neighbour George Sand. The artist's workshops, garden and tearoom are beyond charming. *vie-romantique.paris. fr; 16 rue Chaptal; closed Mon; free.*

Cimetière du Montparnasse

This 19-hectare cemetery opened in 1824 and is Paris' second largest after Père Lachaise. Famous residents include writer Guy de Maupassant, playwright Samuel Beckett, photographer Man Ray, industrialist André Citroën, Captain Alfred Dreyfus of the infamous Dreyfus Affair, legendary singer Serge Gainsbourg and philosopher-writers Jean-Paul Sartre and Simone de Beauvoir. *paris.fr; 3 bd Edgar Quinet; free.*

Parc Monceau

Marked by a neoclassical rotunda at its main entrance, beautiful Parc Monceau was laid out by Louis Carrogis Carmontelle in 1778–79 in what was then called the 'English style' with winding paths, ponds and flower beds. An Egyptian-style pyramid is the only original folly remaining today, but there's a bridge modelled after Venice's Rialto, a Renaissance arch and a Corinthian colonnade. *paris.fr/equipements/parc-monceau-1804; 35 bd de Courcelles; free.*

FIRST SUNDAYS OF THE MONTH

Many of Paris' top museums are free on the first Sunday of every month: Centre Pompidou (centrepompidou. fr; place Georges Pompidou); Musée d'Orsay (musee-orsay.fr; rue de Lille); Musée Rodin (musee-rodin.fr; 77 rue de Varenne); and Musée de l'Orangerie (musee-orangerie. fr; Jardin des Tuileries), to name just a few. The Louvre (louvre. fr; rue de Rivoli), Arc de Triomphe (monuments-nationaux.fr; place Charles de Gaulle) and Château de Versailles (chateauversailles. fr; Versailles) have free Sundays on some months only.

EUROPE

VENICE

Europe's elite come to Venice to lounge in lavish hotels and buzz around in velvet-padded gondolas. But some of the finest experiences cost nothing at all: seeing St Mark's Basilica is gratis, and the streets and palaces look best reflected for free in the Grand Canal.

✋ Basilica di San Marco

Venice's byzantine basilica is the apotheosis of the city's self-invention. It took 800 years to build and wrap the bones of St Mark the Evangelist in a golden carapace of mosaics, resulting in this magnificent monument on Piazza San Marco. This is all yours for free, although you can skip the queue by pre-booking (€3). *basilicasanmarco.it; Piazza San Marco; free.*

✋ DIY architecture tour

Forget the museums. The real beauty of Venice is the city itself: an architectural masterpiece of floating palaces in Istrian stone. Since the 7th century, the city's signature flair has evolved into a dazzling composite of styles. Jump on vaporetto 1 (every 10 minutes; €7.50) for a zigzag cruise through a thousand years of history down the Grand Canal to San Marco and the Lido.

✋ Cimitero di San Michele

Surrounded by a high wall and tall cypresses, and presided over by Codussi's 1469 pearly white Renaissance church, San Michele is Venice's island of the dead. Ezra Pound, Sergei Diaghilev and Igor Stravinsky all lie here, and music-lovers and architecture buffs come to pay their respects and admire the basalt-clad bunker built by David Chipperfield. *Free.*

✋ Priceless views for free

Entering Venice's *palazzi* comes with a hefty price tag, but there's no charge for the views from the myriad bridges across the city's canals. The Rialto Bridge attracts the hordes, but the views are just as good, and quieter, from the wooden Ponte dell'Accademia. *Free*

✋ Rialto market

The Rialto market and pescaria (fish market) are a part of Venice's living history. Sustainable fishing practices are not new in this seafood emporium: marble plaques show regulations set centuries ago for the minimum allowable size for catches. It's like a dry-dock aquarium tour of the lagoon's sea creatures. *Campo della Pescaria; free.*

Below: exploring markets such as the Rialto and its associated farmers' market is a great way to learn about local produce.

© FooTToo | Shutterstock

EUROPE

ISTANBUL

With one foot in Asia and the other in Europe, İstanbul is a cultural whirlpool. Meandering merchants and bargain-hunting travellers have hustled and haggled in the bazaars for centuries, and there's plenty to delight even when your budget's tight.

Blue Mosque

The stunning 400-year-old Sultanahmet Camii was built by Sultan Ahmet I and nicknamed for the 20,000 blue tiles that flow across its cascading domes. Avoid prayer time, which happens five times a day (listen for the *ezan* being chanted from the six minarets). *sultanahmetcamii.org; closed during prayers; free.*

Dancing dervishes

Listed by Unesco, the Sema ceremony is performed by the Contemporary Lovers of Mevlana each Thursday evening. After watching the dancers (often called whirling dervishes) you can speak with the mystics to find out more. *emav. org/emav; Silivrikapı Mevlâna Cultural Center, Mevlânakapı Mah; Thu; free.*

Doğançay Museum

This atmospheric modern art museum displays a cool collection of work by Burhan and Adil Doğançay, two of Turkey's most important painters. Five decades of artistic expression is exhibited; visit early afternoon to score a free Turkish tea. *dogancaymuseum.org; Balo Sokak 42; closed Mon; free.*

Grand Bazaar

It's free to visit Istanbul's legendary public market, and legions of shoppers take full advantage, thronging the cool, covered alleys. Travellers have been rubbing lamps and haggling for bargains here for 600 years, and it's not unusual to receive a free cup of tea or a cube of Turkish delight from enthusiastic vendors. *Kalpakçılar Cd; closed Sun; admission free.*

Elgiz Museum

Turkey's original contemporary art museum features a privately owned collection which includes an impressive selection of works by international names like Tracey Emin and Jan Fabre, and local talent including Ömer Uluç and Princess Fahrelnissa Zeid. *elgizmuseum. org; Beybi Giz Plaza; closed Sun & Mon, by appointment Tue; free.*

On the Eastside

A Bosphorus boat-crossing is an essential experience, and the enigmatic east bank offers plenty to see for not that much. Have a nose at Beylerbeyi Palace (former summerhouse of the sultans; 40TL) and the 16th-century İskele Camii (aka Mosque of the Pier and Mihrimah Sultan Mosque) in Üsküdar, commissioned by Süleyman the Magnificent's daughter. *Free.*

EUROPE'S BEST FREE MUSEUMS AND GALLERIES

Museum-mooching can be a pricey pursuit, but there's no need to dust the cobwebs off your wallet to enjoy these esteemed collections.

EUROPE

THE PRADO, SPAIN

Formerly a royal collection, this feast of fine art is now free every evening. One of the world's greatest galleries, it boasts paintings by Goya, Raphael, Rubens and more. *museodelprado. es; Paseo del Prado, Madrid, Spain; free evenings Mon-Sun.*

02 SCHUTTERSGALERIJ, NETHERLANDS

One of the world's only 'museum streets', this walk-through gallery at the Amsterdam Museum mixes historic portraits with a sprinkling of Dutch Masters, some amusing modern artwork and a 350-year-old statue of Goliath. *amsterdammuseum.nl; Kalverstraat 92, Amsterdam, Netherlands; free.*

BERLIN WALL MEMORIAL, GERMANY

This poignant museum preserves the last surviving piece of the Berlin Wall, tells the story of the ghost station of Nordbahnhof S-Bahn, and relates tales of horror, heroism and hope on Bernauer Strasse. *berlin er-mauer-gedenkstaette.de; Bernauer Strasse 119, Berlin, Germany; free.*

VATICAN MUSEUMS, ROME

Founded by Pope Julius II in the early 16th century, the Vatican Museums boast one of the world's greatest art collections, alongside the epic Sistine Chapel. Even more amazingly, it's free to visit one Sunday per month, though you'll have to come early to beat the crowds. *m.museivaticani.va; Vatican City; free last Sun of month.*

03 BRISTOL STREET ART, ENGLAND

Free-range chin-strokers can enjoy some of the world's best modern street art in Bristol. In Banksy's home town, the al-fresco scene is sensational, with public spaces splattered with murals and installations by guerrilla artists. *bristol-street-art.co.uk; Bristol, England; free.*

MUSÉE CARNAVALET, FRANCE

History meets art in this peculiarly Parisian institution, where 100 Tardis-like rooms transport visitors through time, via paintings and artefacts from the City of Light's illuminating past. *carnavalet.paris.fr; 16 Rue des Francs-Bourgeois, Paris, France; closed Mon; free.*

ROYAL DANISH ARSENAL MUSEUM, DENMARK

This ensemble ranges from samurai swords to a WWII German V-1 flying bomb, via various guns and spiky things. The cache was begun in 1604, by King Christian IV. *natmus.dk; Tøjhusgade 3, Copenhagen, Denmark; closed Mon; free.*

04 BRITISH MUSEUM, ENGLAND

London's full of famous freebies, but this mothership museum, packed with souvenirs accumulated during Britain's years of colonisation, is top. You'll never get around in one day – just come back. *britishmuseum.org; Great Russell St, London, England; free*

MOSCOW

While Moscow has a long-held reputation as an oligarch's playground, it's not all bad news for budget travellers. Keep the spending in check with plenty of green spaces, epic outdoor monuments, markets, free museums and cheap Russian sustenance.

✋ Cathedral of Christ the Saviour

The shining gold domes of the Cathedral of Christ the Saviour are spectacular to admire from outside and even more impressive inside. It's free to stand in awe, neck craned to take in the fresco-covered dome interior. And don't miss *Christ Not Painted by Hand* by Sorokin in the ground-level chapel. *xxc.ru; ul Volkhonka 15; free.*

✋ Lenin's Mausoleum

All it costs to see Lenin's Mausoleum is time spent queueing. The embalmed revolutionary leader has been lying in this tomb at the base of the Kremlin Wall since 1924, mourned daily by loyal Communists. *lenin.ru; Red Square; closed Fri; free.*

✋ Moscow metro art

Riding Moscow's metro not only transports you from A to B but also into a world of opulence, history and Soviet iconography. For the price of a metro ticket, you can tour the stations and take in everything from the Art Deco central hall of Mayakovskaya to the mosaics of military heroes at Komsomolskaya. *Tickets from 57R.*

✋ Novodevichy Cemetery

Pay nothing but your respects to some of Russia's biggest wigs at Novodevichy Cemetery. Wander past tombs for luminaries such as Chekhov, Bulgakov and Stanislavsky, plus more controversial figures including Stalin's wife. *Luzhnetsky pr 2; free.*

✋ Red Square

Red Square lives up to expectations as you enter its vast expanse and your eyes try to take in every bit of architectural wonder, from the scale of the red-brick Kremlin walls to the candy-coloured onion domes of St Basil's cathedral. *Krasnaya pl; free.*

✋ Moscow Tchaikovsky Conservatory

Concert tickets at this musical institution are costly, but you can time a visit to coincide with some of the free performances that are held periodically in the conservatory. Check the website to find gratis concerts. *mosconsv.ru; ul Bolshaya Nikitskaya 13; free.*

Below: Red Square in Moscow offers astounding architecture.

© Helen Filatova | Shutterstock

EUROPE

BERLIN, SAXONY & BRANDENBURG

Berlin is a city that throws itself wholeheartedly into the love of life. Even 30 years after the fall of the Wall, the German capital matches culture and vibrancy with affordable prices, and there's more to see in the nearby regions of Saxony and Brandenburg.

✋ East Side Gallery

It's ironic that Berlin's must-see tourist attraction no longer exists. Thankfully there's the East Side Gallery, a 1.3km-long fragment of the Berlin Wall that's been smothered in street art to become the longest open-air gallery in the world. *eastsidegallery-berlin.de; Mühlenstrasse btwn Oberbaumbrücke & Ostbahnhof; free.*

✋ Holocaust Memorial

Feel the presence of uncounted souls as you meander through the maze that is Berlin's Memorial to the Murdered Jews of Europe. The vast site consists of 2711 concrete slabs of varying height, arranged in a claustrophobic grid that you are free to explore at will. *stiftung-denkmal.de; Cora-Berliner-Strasse 1; free.*

✋ Sachsenhausen Memorial and Museum

Tens of thousands of prisoners perished behind the sinister gates of the Sachsenhausen Concentration Camp between 1936 and 1945. Now a memorial, gripping exhibits in sites like the infirmary barracks and the execution area keep haunting memories alive. *sachsenhausen-sbg.de; Strasse der Nationen 22, Oranienburg; free.*

✋ Lunchtime concert at the Philharmoniker

Instead of stuffing your tummy, why not feed your ears with a lunchtime foyer concert played by the *meister* musicians of the Berliner Philharmoniker? Weekly performances lure crowds of culture-hungry fans, so arrive early to stake out a spot. *berliner-philharmoniker.de; Herbert-von-Karajan-Strasse 1; Wed Sep-Jun; free.*

✋ Stay at the Bauhaus

It's not every day that you get to stay in the home of an art movement, but in Dessau, an hour and a half south of Berlin, the former studios used by the artists and designers of the Bauhaus have been converted into surprisingly affordable accommodation. *bauhaus-dessau.de; Gropiussallee 38; single/double from €40/60*

✋ Priceless Prussia without the pricetag in Potsdam

Potsdam, just 25km southwest of Berlin, was the former Prussian royal seat, and it overflows with lavish monuments thrown up by Frederick the Great. You'll pay to enter many palaces, but there's no charge to explore Sanssouci Park, whose lavish gardens border Sanssouci Palace and the New Palace.

BARCELONA

Barcelona is the perfect mix of seaside resort and cosmopolitan city. With eternal good weather and a reputation for partying till dawn, this is a place to admire Modernist architecture and sun-bask by day, then graze inexpensive gastronomy by night.

Museu Picasso
This museum provides an insightful look into the formative years of Picasso's early life. It's well worth a visit to see the evolution of a budding genius (via more than 3500 works) but get there early. *museupicasso.bcn.cat; Carrer Montcada 15-23; closed Mon; free first Sun of the month & every Thu after 4pm.*

Modernist architecture
The town is littered with impressive works by Modernist greats such as Gaudí, Domènech i Montaner and Puig i Cadafalch, whose enchanting fairytale facades can be seen and revered from street level. The highest concentration of these works is in the Eixample district, in the area known as the Quadrat d'Or (the 'Golden Square'). *Free.*

Centre de Cultura Contemporània de Barcelona
An innovative and eye-catching creative space, the CCCB offers everything from exhibitions and film nights to lectures and debates, along a general theme of urbanisation and contemporary culture. A visit is always thought-provoking. *cccb. org; Carrer Montalegre 5; closed Mon; free Sun afternoon.*

El Born Centre Cultural
El Born is a landmark of Catalan Modernism. Perched over exposed subterranean ruins, it's a politically-charged testament to the destruction wrought by Spanish King Philip V after Barcelona's defeat in the War of the Spanish Succession. *elborncentrecultural. barcelona.cat; Plaça Comercial 12, closed Mon; free, admission to exhibitions from €3.*

Public/Street art
Barcelona's wealth of public art is gratis to explore. Seek out Fernando Botero's fat cat on Rambla del Raval; *Peix*, the giant stainless-steel fish on the waterfront designed by Frank Gehry; Roy Lichtenstein's iconic 15m-high *El Cap de Barcelona* at Port Vell; and Miró's *Woman and Bird* sculpture in Parc de Joan Miró. *Free.*

Flicks al fresco
Barcelona's balmy weather makes for perfect outdoor movie viewing. Try the scenic square of CCCB (Gandules Outdoor Cinema Season; cccb.org) or the open beaches of the Catalan coast (Cinema Lliure a la Platja; cinemalliure. com; Jul & Aug). Movies can be slightly hit-and-miss but are usually conversation-starters. *Free.*

DUBLIN

In Ireland's capital, high culture and light-hearted craic go hand-in-hand, history is literally etched into every corner, and all this often comes for free. In the legendary pubs, the iconic black stout runs freely...but rarely cheaply.

National Museum of Ireland
Explore Ireland's heritage via four million objects. The three Dublin venues tackle archaeology (Kildare St; from prehistory to the Viking Age); decorative arts and history (Collins Barrack); and Natural History (Merrion St). *museum.ie; closed Mon; free.*

Trinity College Dublin
It's free to swan around the beautiful green grounds and cobbled squares of Trinity College, which dates back to 1592. Former students include Oscar Wilde, Samuel Beckett, Jonathan Swift and, er, Courtney Love. *tcd.ie; College Green; free.*

Bank of Ireland/Irish Parliament House
Behind a bank facade, this 1733 Palladian pile was once the world's first purpose-built parliament building, and the inspiration for Washington's House of Representatives. The British insisted the interior be altered to deter future notions of independence; ask to enter the ornate chamber. *College Green; weekdays; free.*

Irish Museum of Modern Art
The building that houses Ireland's premier collection of modern art was based on Les Invalides in Paris, which explains the formal facade and courtyard. View 3500 artworks by Irish and global artists. *imma.ie; Royal Hospital, Kilmainham; closed Mon; free.*

National Gallery of Ireland
Packed to the rafters with Irish art and works from all the major European schools, this oasis of culture and calm in the midst of Georgian Dublin has 15,000 paintings and sculptures to feast your eyes on. Ask to borrow audioguides and drawing kits. *nationalgallery.ie; Merrion Sq; free.*

Sunday concerts
Sundays @ Noon gigs at The Hugh Lane Gallery kicked off in 1975, and the space still fills with the classical sounds of Irish and international musicians. The gallery itself (free) houses Francis Bacon's studio plus some surprising contemporary art. *hughlane.ie; Parnell Sq; concerts Sep-Jun; free.*

Below: the National Gallery of Ireland will occupy a day of any visit to Dublin.

EUROPE

© Benoit Daoust | Shutterstock

STINGE HENGE
ANCIENT STONES FOR FREE

If the gift shop, fences and entry fees at Stonehenge leave you stone cold, explore these free-to-see rock stars instead.

POULNABRONE DOLMEN, IRELAND

Balanced on the breathtakingly bleak Burren along Ireland's western flank, this elegantly poised dolmen (portal tomb) is 5000 to 6000 years old. Although there are half-hearted ropes (to discourage half-hearted climbers) you can walk freely around it. *County Clare, Ireland; free.*

TRETHEVY QUOIT, ENGLAND

Perched on a promontory overlooking the confluence of two streams, this quoit (megalithic tomb) was built between 3700 BCE and 3300 BCE, probably to house bones – although the acidity in the earth has long since eaten the evidence. *english-heritage.org.uk; Cornwall, England; free.*

DOLMEN DE MANÉ KERIONED, FRANCE

Carnac is covered with hundreds of standing stones and dolmens. The main site charges €9 for tours (except in winter, when access is free), but nearby Dolmen de Mané Kerioned (three dolmens and small menhirs, all dating to c 3500 BCE) is free to all year-round. *carnactourism.co.uk; Carnac, France; free.*

DOLMEN DE MENGA, SPAIN

A behemoth burial mound, the chamber of this 25m-long tumulus – built around 3000 BCE with 32 megaliths – was found to contain hundreds of skeletons when it was opened in the 19th century. *andalucia.org/en/antequera-cultural-tourism-dolmen-de-menga; Antequera, Spain; free.*

CALANAIS STONE CIRCLE, SCOTLAND

This remote string of stone sentinels is sited on a wild and secluded promontory overlooking Loch Roag. Thirteen large stones of beautifully banded gneiss are arranged, as if in worship, around a 4.5m-tall central monolith. *historic environment.scot; Isle of Lewis, Scotland; free*

POSKÆR STENHUS, DENMARK

Denmark's largest surviving round barrow/henge boasts 23 stones, including an 11.5-tonne capstone, which has stood on its neighbours' shoulders since it was balanced there around 3300 BCE. Look through the Poskær burial chamber at dawn on the spring equinox to see the sunrise. *visit djursland.com; Knebel, Denmark; free.*

RING OF BRODGAR, SCOTLAND

Probably Britain's best-preserved stone circle, this 4000-year-old site retains 27 of the original 60 megaliths erected here. Measuring 104m across, its purpose remains an enigma – used for astronomy possibly, or religious rituals. *historic-scotland.gov.uk; Orkney, Scotland; free*

BRYN CELLI DDU, WALES

The 'Mound in the Dark Grove' is a 5000-year-old Neolithic chambered tomb and henge, designed to precisely align with the rising sun during the summer solstice, when a shaft of light penetrates the passageway and illuminates the inner burial chamber.

FOOD & DRINK

Europe's food is a feast as varied as the continent itself. Italy, France and Spain stand proudly in the line-up of top cuisines in the world, while other gastronomic hubs – Georgia, Cyprus, Denmark, Estonia, Slovenia – slip quietly under the radar, waiting to be discovered.

How much you pay for a good meal can vary immensely. In Iceland, Sweden, Denmark and Norway, a petrol-station hot dog can cost as much as a restaurant main course in Bulgaria or Romania. But as each country has its cheap feasts, a full stomach and full wallet is not an unachievable goal.

Above: eating out in Europe. Right: take-out snacks such as these panini in Venice will keep you going all day.

STREET FOOD TIPS

Street food has seen a major revival in Western Europe, perhaps kickstarted by by travellers bringing back a taste for the exotic from their travels. Most food markets and many tourist hangouts have gatherings of inexpensive, globe-trotting food stalls serving a peripatetic menu of curries, stir-fries, tortillas and more.

Along with the international snacks, European countries have their own indigenous portable food traditions too: pasties and pies in Britain; *knish* (stuffed buns) in Eastern Europe; *arancini* (fried rice balls) in Italy; *börek* (stuffed filo pastries) in Turkey, all served up by bakeries and delis everywhere.

SIT-DOWN DINING

Restaurant dining will ramp up your expenses, but it doesn't have to empty your bank account. Learn the cheap times and places to eat in each location, and prudent gourmands will feast like a knight on a squire's budget.

- **Don't overlook bakeries** – with a side of coffee, a fresh croissant, focaccia or pretzel can be all it takes to set you up for the day.
- **Avoid tourist hubs** – restaurants in touristy areas always bump up prices; head to a quieter suburb nearby and prices will magically shrink to more reasonable levels.
- **Go big at lunchtime** – across Europe, the cheap lunch is a definite thing. Look for cut-price lunchtime specials everywhere; the choice may be limited, but so are the prices.
- **Set menu feasting** – set menus offer big savings. In France, look for the *table d'hôte*, *prix fixe* or *formule* menu (a starter and main, and sometimes dessert); in Spain, find the *menu-del-dia*; in Germany, look for *das menü* (as opposed to *speisekarte*, the normal menu).
- **Slow down** – Italy's Slow Food movement (slowfood.com) has spread Europe-wide, promoting local foods and produce, produced the traditional way, often at pocket-pleasing prices.

BAR FOOD

Some of Europe's best meals are served not in restaurants but in cafes, pubs and bars. Even the UK, famed for its hard-drinking mentality, has caught the bar-food bug, with pubs serving sliders and Korean tapas in place of the traditional crisps and pork scratchings.

There's no guarantee that bar food will be cheap – indeed, the bill in some can trigger global financial meltdowns – but it needn't blow the lid off your dining budget, particularly in enlightened, foodie corners of the continent.

- **Main course or munchies?** – most watering holes have two tiers of food: light, pocket-sensitive, filling fast food such as chips, burgers and sandwiches,

CUISINE OF EMPIRES

Europe's planet-plundering empires didn't just fatten the coffers back home: they also paved the way for today's vastly enriched gastronomic cultures. Many countries in Europe have an unofficial second national cuisine, prepared with great skill for pocket-pleasing prices by emigres (and their descendants) from former colonies. In the UK, Indian food is king – but that's misleading, as 87% of Indian restaurants are actually run by cooks from the Sylhet region in Bangladesh. In France, it's all about Vietnamese noodles and Moroccan tagines; in the Netherlands, Indonesian satay rules the roost. Europe has plenty to thank its empires for.

EUROPE

alongside proper knife-and-fork main courses, at much higher prices.

- **Bar snack suppers** – bar food in Europe means more than peanuts on the bar. In Italy, *apertivo* bars lay on plates of (usually delicious) small snacks to munch with drinks, sometimes for free.
- **Spanish delights** – small snacking tapas plates are a cheap way to dine; in the Basque region, *pintxo* bars line up snacking morsels on the bar from as little as €1.

FOOD MARKETS

We've all seen footage of celebrity chefs getting giddy as children when confronted by the quality of ingredients in Europe's produce-piled food markets. The further south and east you go, the more open-air markets are melded into public life; by the time you reach the Med, al fresco markets do more business than indoor superstores. There's normally a section of every market set aside for food stands, so look for the people with paper plates and contented expressions.

- **Come early or late** – traders often toss in freebies for the first customers of the day, or offer cut-price stock to clear the decks before packing up for the evening.
- **Be a sampler** – many market vendors will let you try before you buy, even if it's only a morsel. Make a meal from grazing free samples at epicurean bazaars such as London's Borough Market.
- **Know your onions** – some things are just downright expensive (truffles, caviar, pistachios, Modena vinegar). Avoid these immoderate foodstuffs and you'll avoid a high bill for a market munch.

SMART DRINKING

Alcohol prices are creeping up across Europe as governments strive to keep a

Top left: a fruit stall in Moscow. Bottom left: a food truck in Sweden. Above: open-air cafes in Amsterdam.

lid on problem drinking, meaning boozy nights out are rarely a bargain. Watch how locals drink and you'll get a clue to the best techniques for economical merriment.

• **Drink like a local** – in some places, house wine flows like water, while spirits are reserved for red-letter days; in other countries, beer is cheap and wine prohibitive. In Russia, beer was only recently reclassified as an alcoholic drink subject to alcohol taxes. Most countries have a cheap firewater (*pastis* in France, schnapps in Germanic countries, *rakia* in Eastern Europe) that's ideal for budget sessions, though it can be friendlier on the pocket than the palate.

• **Quaff coffee instead** – Europe has a rich tradition of meeting up for a drink with no alcohol involved. Tea or coffee often come with a chocolate or biscuit to deliver a sugar rush as well as a caffeine buzz.

• **Take it away** – take-outs are invariably cheaper; just make sure you stay on the right side of public drinking laws (alcohol on public transport is usually a no-no).

HIGHLIGHTS

MERCATO CENTRALE, FLORENCE
Florence's oldest food market is where the feast begins. Admire the produce and ingredients downstairs, then head up to the first floor for artisan food stalls serving Tuscany's favourite foods: steaks, pizza, gelato, pasta and – for the brave – tripe panini. *p130*

LADURÉE, PARIS
A patisserie with a pedigree lasting a century-and-a-half, Ladurée created the very first macaron in the 1930s and never looked back. It's set amid the glamour of the Champs Élysées but even budget travellers can afford to sample the signature treats. *p132*

BERGARA BAR, SAN SEBASTIÁN
San Sebastián is *pintxos* (bar snack) central, and this bar is one of the best places to graze, with everything from *chupito* (spider crab mousse) to anchovy tortillas, washed down with excellent local wines and beers. Yum! *p138*

GASOLINE GRILL, COPENHAGEN
Even pricey Denmark has its bargains, and this petrol station-turned-burger house proves the point. Meat and veggie burgers are wrapped in brioche buns and dished out with chips, dips and desserts. Ask for extra napkins; you'll need them. *p135*

EUROPE

FLORENCE & TUSCANY

Tourist hordes push up prices at Florence's museums and palazzi; get payback by feasting economically at small hole-in-the-wall eateries. The city is also the gateway to Tuscany, where the Italian idyll of art, scenery and slow food comes vividly to life.

✋ Mercato Centrale, Florence

Florence's oldest and largest food market is a maze of stalls beneath a striking iron-and-glass canopy from 1874. Head to the first floor's buzzing food hall with its artisan stalls serving steaks, burgers, tripe panini, vegetarian dishes, pizza, gelato, pastries and pasta. *mercatocentrale.it; Piazza del Mercato Centrale; closed Sun; dishes from €5.*

✋ Mariano, Florence

A Florentine favourite for its simplicity and pleasingly low prices, around since 1973. From sunrise to sunset, this brick-vaulted, 13th-century cellar gently buzzes with locals propped at the counter sipping coffee and wine or eating salads and panini. *Via del Parione 19r; closed Sun; panini from €3.50.*

✋ Osteria Il Buongustai, Florence

Run with breathtaking speed and grace by Laura and Lucia, and a surprise find in the touristy centre, 'The Gourmand' is unmissable. Lunchtimes heave with locals and savvy students who flock here to fill up on tasty Tuscan home cooking at a snip of other restaurant prices. *Via dei Cerchi 15r; closed Sun; meals from €15.*

✋ Trattoria Mario, Florence

Arrive by noon to ensure a spot at this noisy, busy, brilliant trattoria – a legend that retains its soul (and allure with locals) despite being in every guidebook. Charming Fabio, whose grandfather opened the place in 1953, is front of house while big brother Romeo and nephew Francesco cook with speed in the kitchen. *trattoriamario.com; Via Rosina 2r; closed Sun; meals €25.*

✋ La Toraia, Florence

Find this cherry-red artisan food truck, parked a 15-minute stroll east of Florence's Piazza di Santa Croce. Sweet 140g

Below: a queue for gelato in Florence – a line of locals is often a good way to spot a popular parlour.

burgers, crafted from tender Chianina meat sourced at the family farm, are topped with melted pecorino cheese. *latoraia.com; Lungarno del Tempio; mid-Apr–mid-Oct; meals from €5.*

👋 Terrazza Menoni, Florence
Luca Menoni's meat stall inside Florence's Sant'Ambrogio covered market has been a favourite with locals since 1921, and now the family has opened a zero-kilometre *risto macelleria* (butcher's eatery) upstairs. Everything is homemade and ingredients are sourced fresh from the morning market. *terrazzamenoni.it; Piazza Ghiberti 11; closed Sun; meals from €15.*

👋 Gelateria Pasticceria Badiani, Florence
Known across Italy for the quality of its handmade gelato and pastries, Badiani is set in the Campo di Marte neighbourhood just outside the centre, but it is – as any local will tell you – worth the walk. The house speciality is Buontalenti gelato, a creamy concoction with flavourings that remain secret. *gelateriabadiani.it; Viale dei Mille 20r; gelato from €3*

👋 #Raw, Florence
Should you desire a turmeric, ginger or aloe vera shot or a gently warmed, raw vegan burger, innovative #Raw hits the spot. Everything served here is freshly made and raw – to sensational effect. Herbs are grown in the biodynamic greenhouse of charismatic and hugely knowledgeable chef Caroline. *hashtagraw.it; Via Sant'Agostino 11r; closed Mon; meals €8-15.*

👋 Antica Torteria Al Mercato da Gagarin, Livorno
There is no finer taste of old-world Livorno than this retro snack bar. Push through the plastic fly-net veiling the unmarked door and order the house speciality: a sensational *cinque e cinque* sandwich stuffed with scrumptious *torta di ceci* (chickpea pancake, fried in peanut oil and salted). *Via del Cardinale 24; closed Sun; sandwich €2.50-3.20.*

👋 Sergio Falaschi, San Miniato
The most famous *macelleria* (butcher's shop) in San Miniato, run by the same family since 1925, is the place to come for carnivorous cuisine at (for Tuscany) modest prices. Head to the back to feast on feisty dishes made from local Chianina beef and *cinta senese* (Siena pork). *sergiofalaschi.com; Via Augusto Conti 18-20; meals from €25*

👋 Il Cedro, Moggiona
The cuisine of the Casentino valley is the drawcard of this charming, family run village bistro, squirrelled away for the last 45 years in the tiny hamlet of Moggiona, on the winding road to Camaldoli. There is no menu – rather, seasonal, traditional dishes of the day are chalked on the board. *ristoranteilcedro.com; Moggiona; closed Mon; meals €24.*

👋 Gelateria Dondoli, San Gimignano
Think of it less as ice cream, more as art. Former gelato world champion Sergio Dondoli is famed for creations including Crema di Santa Fina (saffron cream) gelato and Vernaccia sorbet. *gelateriadipiazza.com; Piazza della Cisterna 4; gelato from €2.50.*

👋 Antica Osteria Dell'Agania, Arezzo
Operated by the Ludovichi family since 1905, Dell'Agania serves diehard traditional fare, the cornerstone of Tuscan dining. Specialties include sensational antipasti, homemade pasta and *secondi* ranging from snails to lambs' cheeks. *agania.it; Via Mazzini 10; closed Mon; meals €20.*

EUROPE

FREE FLORENCE

Having filled up on fabulous Florentine food, work off the calories by strolling around the city's abundant free sights. The Piazza del Duomo is flanked by the wedding cake-like Cathedral of Santa Maria del Fiore (museumflorence. com; free) and you can view one of the finest Last Supper paintings at Cenacolo di Sant'Apollonia (polomuseale-toscana. beniculturali.it; Via XXVII Aprile; free). Across the river is fresco-filled San Miniato al Monte (sanminiato-almonte.it; Via delle Porte Sante; free).

PARIS

The City of Love has a similar attraction to money as it does for lovers. To stop pockets emptying at an alarming rate, plan mealtimes ahead to take advantage of the city's best bargain bistros, food markets and lip-smacking patisseries.

Marché d'Aligre

Browsing is free at this lively market, but come at lunchtime to splurge on the fabulous foodstuffs piled on all sides. From seafood to cheese and everything in between, it offers hours of delicious entertainment and colourful photo opportunities while you snack. *marchedaligre.free.fr; rue d'Aligre; closed Mon.*

Ladurée

One of Paris' oldest patisseries, Ladurée has been around since 1862 and first created the lighter-than-air, ganache-filled macaron in the 1930s. Its tearoom is the classiest spot on the Champs to indulge in macarons, pastries or more formal meals (at steeper prices). *laduree.fr; 75 av des Champs-Élysées; pastries from €2.60.*

Café de la Nouvelle Mairie

Shhhh...just around the corner from the Panthéon but hidden away on a small, fountained square, this cafe-restaurant and wine bar is a tip-top neighbourhood secret, serving organic wines and delicious seasonal bistro fare, from oysters and ribs (à la française) to grilled lamb sausage over lentils. *19 rue des Fossés St-Jacques; closed Sat & Sun; mains from €8.*

Jacques Genin

Wildly creative chocolatier Jacques Genin is famed for his flavoured caramels, *pâtes de fruits* (fruit jellies) and exquisitely embossed *bonbons de chocolat* (chocolate sweets). But what completely steals the show at his elegant chocolate showroom is the salon de dégustation (aka tearoom), where you can order a pot of outrageously thick hot chocolate and legendary Genin millefeuille, assembled to order. *jacquesgenin.fr; 133 rue de Turenne; closed Mon; deluxe pastries from €9.*

Breizh Café

Be it a simple sweet crêpe smeared with melted hand-churned Breton butter, a *galette* (savoury pancake) stuffed

Below: there are numerous covered markets around Paris and they're a great place to pick up cheap snacks.

EUROPE

with smoked duck breast, pan-fried mushrooms and Comté cheese, or a St-Malo seaweed-and-egg-filled crêpe roll, the Batignolles outpost of France's roaringly popular Breizh crêperie does not disappoint. *breizhcafe.com; 31 rue des Batignolles; crêpes from €6.80.*

🖐 Holybelly

Friendly vibes, sassy breakfast 'n' lunch dishes and specialist coffee define these twin cafes at No 5 (all-day pancakes and eggs) and No 19 (seasonal dishes to share). Be it Holybelly's signature bacon cured in organic maple syrup, kasha porridge with poached pear, or winter asparagus with hazelnut dukkah, dining here is never dull. *holybellycafe.com; 5 & 19 rue Lucien Sampaix; dishes from €6.*

🖐 Du Pain et des Idées

This traditional bakery with an exquisite 19th-century interior is famed for its naturally leavened bread, orange-blossom brioche and flavoured *escargots* (scroll-like 'snails'). Its mini savoury *pavés* (breads) flavoured with Reblochon cheese and fig, or goat's cheese, sesame and honey, are perfect for lunch on the run. *dupainetdesidees.com; 34 rue Yves Toudic; closed Sat & Sun; breads from €1.20.*

🖐 L'Avant-Poste

'Eco-responsible' is the tasty buzzword at this foodie outpost, set in a former wig shop in the earthy 10e neighbourhood. Fresh, seasonal produce is sourced from artisan producers and the menu changes daily. Tuck into seaweed-spiked beetroot borscht or Corsican *panzetta* (pork) with an old-fashioned mix of spelt, mushrooms, courgette and chard. Reservations vital. *lavantposteparis.fr; 7 rue de la Fidélité; closed Sun & Mon; two-course menu €16.50.*

🖐 Papilles

No spot in northern Paris satisfies brunch cravings like this hipster coffee house with exposed red brick, velour seating and a menu signed off by Franco-Vietnamese chef Céline Pham. The brunch menu is served all day and fabulously mixes French, Vietnamese and world flavours in dishes such as *bánhnini* (a marinated chicken, coriander and pickle toasted sandwich). *papilles-restaurant.com; 77 rue de Rochechouart; mains from €7.50.*

🖐 Berthillon

Founded here in 1954, this esteemed *glacier* (ice-cream maker) is still run by the same family today. Its 70-plus, all-natural, chemical-free flavours include fruit sorbets (pink grapefruit, raspberry and rose) and richer ice creams made from salted caramel, candied Ardèche chestnuts, Armagnac and prunes and more. *berthillon.fr; 29-31 rue St-Louis en l'Île; closed Mon & Tue; ice cream from €3.*

🖐 Madito

Teensy Madito prepares superb Lebanese cuisine daily from scratch. With just 20 seats, book ahead to feast on starters such as *makdous* (aubergine stuffed with red peppers and walnuts) or *warak enab* (vine-leaf-wrapped rice and spiced beef), followed by *tawouk* (lemon- and yoghurt-marinated chicken). *madito.fr; 38 rue de Citeaux; closed Sun & Mon; mains from €11.*

🖐 Picnic in Jardin du Luxembourg

With its Renaissance gardens, impressive statuary and status as the home of France's Senate, Jardin du Luxembourg is Paris' grandest public park. But for all that magnificence, locals use it for down-to-earth picnics. Pack a basket and enjoy the most glamorous sandwich setting of your life. *Rue de Vaugirard; free.*

BROCANTES AND VIDE-GRENIERS

Every weekend, ephemeral little neighbourhood markets pop up all over Paris. For *vide-greniers* (literally 'empty attics'), entire streets close to traffic so residents can put their unwanted wares on show for a crowd of eager rummagers. It's a yard sale Paris-style, with a festival atmosphere, live music (sometimes) and plenty of food (always). *Brocantes* are more formal affairs, with official antique-sellers setting up displays of lovely precious old things in covered markets – heaven for browsers. You can find a daily agenda for both at brocabrac.fr. *Admission free.*

EUROPE

ROME

Rome can be surprisingly inexpensive for a European capital, thanks in part to the abundance of free-to-access ancient historical sites. When it comes to food, you can pay the earth, or dine deliciously and economically at €3-per-slice pizzerias.

🖐 Freni e Frizioni bar

Once a mechanic's workshop (hence the name 'Brakes and Clutches'), this hip bar draws bright young things who like to sip well-priced cocktails and nibble *aperitivo* snacks in the piazza on a warm evening. *freniefrizioni.com; Via del Politeama 4-6; aperitivo from €6.*

🖐 La Renella

To slurp down pizza in a charming Roman neighbourhood filled with cool bars and cobbled streets, head to Trastevere's La Renella. Compared to a sit-down meal, it's a steal, and open practically 24 hours. *Via del Moro 15-16; pizza slices from €2.50.*

🖐 Mercato di Campagna Amica al Circo Massimo

Seek out this non-touristy market to sample local olive oil and honey, then grab some lunch items and make for one of the picnic tables or the nearby Circus Maximus. *facebook.com/mercatocircomassimo; Via San Teodoro 74; Sat & Sun, closed Aug.*

🖐 Picnicking at the Villa Borghese

Do as the Romans do and simply stroll through the Villa Borghese's manicured English-style gardens, then lay out a picnic from a local deli. Find good gelato nearby on Viale Regina Margherita. *Via Pinciana; dawn-dusk; free.*

🖐 Fior di Luna

Forget the tourist ice cream parlours; many locals insist that Fior di Luna (Italian for 'Moonshine') makes Rome's best handmade gelato and sorbet. Try the classics or sample inventions such as parmesan and apricot. *fiordiluna.com; Via della Lungaretta 96; closed Mon; gelato from €2.50.*

🖐 Bonci Pizzarium

Gabriele Bonci's *pizzeria a taglio* serves Rome's best pizza by the slice. Scissor-cut squares are topped with seasonal ingredients and served for immediate consumption. There are only a couple of benches so head across to the plaza at the metro station for a seat. *bonci.it; Via della Meloria 43; closed Sun; pizza slices from €5.*

Below: pick up a slice of pizza at Gabriele Bonci's pizzarium.

© Alexandra Bruzzese | Lonely Planet

EUROPE

COPENHAGEN

Copenhagen has a reputation as a wallet-wrecking place to visit – yet it was declared the happiest place on the planet by the Earth Institute. We have a suspicion that well-kept foodie secrets (with fair prices) help keep a smile on locals' faces.

✋ Reffen

This harbourside street-food market is a veritable village of converted shipping containers, peddling sustainable bites from across the globe. Interesting eats include organic polenta, *dosas*, sushi, satay skewers and Filipino BBQ. Drinking options include an outpost of Copenhagen's cult-status microbrewery Mikkeller. *reffen.dk; Refshalevej 167; meals from 80kr.*

✋ Gasoline Grill

Join local gourmands in the line at Gasoline, a petrol station-turned-burger takeaway (there are branches dotted around town). The menu is refreshingly straightforward: four burgers (one vegetarian), fries with a choice of toppings and dips, and two desserts. The meat is organic and freshly ground daily; the buns brioche; and the flavour rich and decadent. *gasolinegrill.com; Landgreven 10; burgers from 75kr.*

✋ La Banchina

This tiny spot serves Danish cafe treats for breakfast, lunch and dinner, cooked beautifully and served with little fanfare. The real magic is the setting, a small harbour cove with picnic tables and a wooden pier, where summertime diners dip their feet while tucking into grub like tender barbecued salmon. *labanchina.dk; Refshalevej 141A; mains from 70kr.*

✋ Café Halvvejen

Cosy, wood-panelled Café Halvvejen channels a fast-fading Copenhagen. The menu is unapologetically hearty, generous and cheap for this part of town, with faithful open sandwiches, *frikadeller* (Danish meatballs) and *pariserbøf* (minced beef steak with egg and onions). *cafehalvvejen.dk; Krystalgade 11; closed Sun; mains from 59kr.*

✋ Brødflov

Wise travellers track down this little organic Frederiksberg bakery for superlative baked goods. The classic Danish *kanelsnegl* (cinnamon roll) and *tebirkes* (poppy-seed pastries) are both wonderfully buttery without being overly rich, while Instagram-worthy tarts come in appetite-piquing flavours like meringue and salted caramel. *broedflov.dk; Falkoner Alle 34; pastries from 20kr.*

✋ Torvehallerne KBH

Food market Torvehallerne KBH is an essential stop on the Copenhagen foodie trail. A delicious ode to the fresh, the tasty and the artisanal, the market's beautiful stalls sell everything from seasonal herbs and berries to smoked meats, seafood, cheeses and *smørrebrød* (open sandwiches). *torvehallernekbh.dk; Frederiksborggade 21; dishes from 55kr.*

KAUPPATORI, FINLAND

Sure it's touristy, but the cobbled square at Helsinki harbour is also the best place in the city to pick up forest-fresh blueberries, crowberries, cloudberries, raspberries, wild strawberries and more bounty from Finland's sprawling woodlands. *myhelsinki. fi/en/see-and-do/sights/ market-square; Helsinki, Finland; closed Sun; free.*

ALBERT CUYPMARKT, NETHERLANDS

Some 260 stalls fill the Albert Cuypmarkt, Amsterdam's largest and busiest market. Stallholders loudly tout their wares while snack vendors tempt passers-by with raw-herring sandwiches, *frites* (fries), *poffertjes* (Dutch pancakes) and caramel syrup-filled *stroopwafels*. *albertcuyp-markt. amsterdam; Amsterdam, Netherlands; closed Sun; free.*

MARCHÉ DES ENFANTS ROUGES, FRANCE

Built in 1615, Paris' oldest covered market is a glorious maze of food stalls selling ready-to-eat dishes from around the globe (Moroccan couscous, Japanese bento boxes and more), as well as produce, cheese and flowers. Meander and dine with locals at communal tables. *Paris, France; closed Mon; free.*

MARKTHALLE NEUN, GERMANY

This delightful 1891 Berlin market hall, with its iron-beam-supported ceiling, hosts local and regional producers and their wares. On Street Food Thursday, a couple of dozen international amateur or semi-pro chefs set up stalls to serve delicious snacks from around the world. There's even an on-site craft brewery, Heidenpeters. *markthalleneun.de; Berlin, Germany; closed Sun; free.*

FOOD-MARKET MAGIC

To enjoy Europe's foodie bounty on a budget, skip the Michelin-starred restaurants and make for the gourmet markets.

BOROUGH MARKET, ENGLAND

Free samples are the order of the day at London's most famous food market, which sprawls through connected courtyards beneath the railway arches at Southwark. Try free cheese, olives and artisan bread chunks dipped in spicy sauces, before buying more substantial street food dishes. *boroughmarket.org. uk; London, England; Wed-Sat; free.*

MARKTHAL, NETHERLANDS

One of Rotterdam's signature buildings, this extraordinary inverted-U-shaped market feels like a giant tube to another dimension. The 40m-high hall is topped by a striking fruit-and-veg-muraled ceiling, and stalls are piled with fine produce and foods to eat on the spot, from spicy Indonesian treats to waffles. *markthal. nl; Rotterdam, Netherlands; free*

MERCAT DE LA BOQUERIA, SPAIN

Barcelona's most central fresh-produce market is one of the greatest sound, smell and colour sensations in Europe. Many of the city's top restaurateurs come here to buy, which vouches for the quality of the market's offerings. Ask to sample morsels like Asturian goat's cheese as you go. *boqueria.barcelona; Barcelona, Spain; closed Sun; free*

NASCHMARKT, AUSTRIA

Vienna's most famous market and eating strip began life as a farmers' market in the 18th century; today it spills an abundance of food treats onto Linke Wienzeile. Come for everything from Austrian schnitzel to Middle Eastern *shakshouka*. *naschmarkt-vienna.com; Vienna, Austria; closed Sun; free.*

SAN SEBASTIÁN

Thanks to the city's Michelin-starred chefs, the name San Sebastián is spoken in hushed tones by food enthusiasts. But you don't need to fork out for expensive meals: aim for value rather than penny prices and you'll feast like a real Euskaldunak (Basque).

Mercado de la Bretxa

Dating to 1870, San Sebastián's original Mercado de la Bretxa has been redeveloped to house flashy chain stores, but adjacent to it is the underground covered market where every chef in the old town comes to get the freshest produce. Stock up on gourmet picnic supplies before heading to Monte Urgull for the sea views. *cclabretxa.com; Plaza la Bretxa; closed Sun; free.*

La Cuchara de San Telmo

This bustling, permanently packed bar offers miniature *nueva cocina vasca* (new Basque cuisine) from a supremely creative kitchen. Unlike many San Sebastián bars, this one doesn't have its *pintxos* laid out; instead, order from the blackboard menu behind the counter. Prices are great and so are the tasty morsels. *C/31 de Agosto 28; closed Mon; pintxos €3–5.*

Bergara Bar

The Bergara Bar is one of Gros' most highly regarded *pintxo* bars and has a mouth-watering array of delights piled on the bar counter, as well as others chalked on the board. You can't go wrong, whether you opt for its anchovy tortilla, *chupito* (spider crab mousse served in a shot glass) or rich foie gras with mango jam. *pinchosbergara.es; C/ General Artetxe 8, Gros; pintxos from €2.50.*

Mala Gissona Beer House

A wooden bar, industrial fixtures and an inviting terrace are the backdrop to a long menu of quality beers on tap. Half are from the house brewery in nearby Oiartzun and the rest from other Basque and international brewers. Soak them up with bar food at fair prices. *malagissona.beer; C/Zabaleta 53.*

Arnoldo Heladería

In business since 1935, Arnoldo Heladería makes San Sebastián's best ice cream – ask anyone. Alongside sorbets such as white peach or mango and passion fruit, they make richer, creamier varieties like sherry-soaked raisin and salted walnut or chocolate-raspberry. *arnoldoheladeria. com; C/Garibai 2; ice cream from €2.30.*

Below: check out the Mercado de la Bretxa, where you can buy all the ingredients for a perfect picnic.

© Mark Read | Lonely Planet

EUROPE

BRUGES

It's easy to see why Bruges is so popular, and you don't need a fat pocketbook to enjoy its history, art and even finer chocolate and beer. Wander the historic, Unesco-listed centre then feast cheaply on moules (mussels) and lambic (naturally fermented beer).

Vismarkt

Going strong since 1821, Bruges' fish market, held on weekdays in a colonnaded arcade, is ideally located for a bit of people-watching, not to mention seafood-browsing. You're sure to work up an appetite, so pop into Den Gouden Karpel cafe (dengoudenkarpel. be) for a takeaway crab sandwich (from €4); grab a bench by the Groenerei Canal and watch the action as you eat. *Steenhouwersdijk; Tue–Fri.*

De Stove

Having just 20 seats keeps De Stove feeling friendly and intimate despite perennially rave reviews. This calm, one-room family restaurant is consistently inventive – fish caught daily is the house speciality, but the monthly changing menu also includes the likes of wild boar fillet on oyster mushrooms, and everything is homemade. *restaurantdestove.be; Kleine St-Amandsstraat 4; closed Wed & Thu; mains from €19.*

Café Vlissinghe

Luminaries have frequented Bruges' oldest pub for 500 years; local legend has it that Rubens once painted an imitation coin on the table here and then did a runner. The interior is gorgeously preserved, with wood panelling and a wood-burning stove. In summer take a beer into the shady garden and play boules. *cafevlissinghe. be; Blekersstraat 2; closed Mon & Tue.*

De Garre

To get the measure of Belgian beer, come to this hidden two-floor *estaminet* (tavern) near the Salvador Dalí museum. Try the fabulous Garre draught, which comes with a thick floral head in a glass that's almost a brandy balloon; the pub will only serve you three of these as they're a head-spinning 11% alcohol. Other great brews include the remarkable Struise Pannepot (€3.50), at a modest 10%. *degarre.be; De Garre 1.*

Markt

The heart of ancient Bruges, the old market square is dominated by the Belfort, Belgium's most famous belfry, but on Wednesday mornings, the attention shifts to a lively food market where locals and visitors mix to buy cheeses, sausages, spit-roasted meat, fruit, vegetables, and authentic waffles at pocket-friendly prices. *the-markt.com; Wed; admission free.*

EUROPE

FESTIVALS & EVENTS

Europe's festivals are some of the most raucous on the planet – literally, in the case of Germany's Rock am Ring (rock-am-ring.com; Mendig; June). However, big events don't come cheap. If you're watching the pennies, look to cultural festivals and oddball local events (cheese-rolling in southern England anyone?).

PRACTICALITIES

Big celebrations in Europe normally mean big bills for rooms and transport. Book well ahead: the early bird gets the worm/

Above: Caribbean culture at London's Notting Hill Carnival. Right: Mauerpark in Berlin is a venue for music festivals.

ticket/Airbnb bed.

• **When in doubt, camp** – multi-day festivals are specifically set up with campers in mind (we apologise for the toilets).

• **Avoid the hols** – Christmas and Easter see prices head into the stratosphere; look for off-season events that won't cost your life savings.

• **Work your way** – festivals always need temporary staff; snag a festival job and you may get your ticket for free.

RELIGIOUS FESTIVALS

Religion dominates the festival calendar. Christmas and Easter see the biggest crowds and biggest prices.

• **Celebrate diversity** – Diwali and other festivals from the Indian subcontinent are some of Europe's biggest parties, adding spectacularly spicy food to the mix.

• **Bibles and boozers** – don't overlook boisterous events such as St Patrick's Day: nominally religious, but the excuse for the biggest party of the year in Ireland.

CARNIVALS

Donning a masquerade mask for the Carnevale di Venezia is like stepping back into the world of Casanova, and Cadiz,

Sitges and the Canary Islands also get in on the carnival action. However, carnivals mean crowds and high prices – book light years ahead to secure accommodation.

• **London's Notting Hill Carnival** – a direct import from the Caribbean, down to the costumes, jerk chicken and bone-shaking sound systems.

• **LGBTQ+ Pride** – Europe-wide parades rival carnivals for extravagance and exuberance.

MUSIC & CULTURE

Big music events across Europe cost big bucks; the best way to save is to plan and pay for everything far in advance.

• **It's not all rock and roll** – pop and rock festivals charge; many literature, art, drama and classical music festivals don't.

• **Fringe events** – even at big-ticket events, look for free shows and performances at smaller stages around the main venues.

• **Make substitutions** – trade Cannes Film Festival (festival-cannes.com; Cannes; May) for the Fusion Film Festival (fusionfilmfestivals.com; months vary), with free screenings in four European cities.

FOOD

Food festivals in Europe are binge-eating banquets, and paid-for events are offset by local delicacies at discount prices. After a day at beer-drenched Oktoberfest in Munich (oktoberfest.de) or Alba's White Truffle Fair (fieradeltartufo.org), you won't need supper.

SPORTS

Big sporting events in Europe offer few bargains. Minor league events and minority sports are cheaper; if you've ever fancied bog snorkelling in Wales, or wife-carrying in Finland, you might even get to participate.

HIGHLIGHTS

FIESTA DE SAN ISIDRO

Madrid's biggest bash, this full-colour fiesta honours the city's patron saint, San Isidro, with endless processions, celebrations and free concerts. Locals put on regional dress for the big day, and the high-octane fun continues through till morning. *p142*

NOTTING HILL CARNIVAL

Multiculturalism is written large at London's biggest street party, where the culture, costumes and creativity of the Caribbean are transplanted to the streets of west London. It's one of Europe's most enthusiastic celebrations, and almost certainly its loudest. *p143*

EUROPE

FESTIVALS

🌐 Carnevale di Venezia, Italy
Venice's annual carnival brings the time of Casanova vividly back to life, with masked balls, powdered wigs and plenty of pageantry. It's an expensive time to be in Venice but, surprisingly, some events are free, and there's no charge to wander the streets snapping photos of masqueraders. *carnevale.venezia.it; Jan–Feb; some free events.*

🌐 Copenhagen jazz festivals, Denmark
Summer's Copenhagen Jazz Festival and winter's Vinterjazz see the Danish capital transform into one huge jazz club, with hundreds of concerts across multiple venues, many of them in the open and for free. *jazz.dk; Jul & Feb; cost varies.*

🌐 Edinburgh's festival frenzy, Scotland
Each August, Edinburgh explodes with events including its Book Festival (edbookfest.co.uk), Military Tattoo (edintattoo.co.uk) and the world-famous, three-week-long Edinburgh Fringe (edfringe.com). Alongside ticketed events, free street entertainment is everywhere.

🌐 Festes de la Mercè, Spain
Barcelona's outrageous celebration is famed for its *castellers* (human towers), *correfoc* (fire-run), and *gigantes*, a parade of effigies of royals and other celebs. *spanish-fiestas.com/festivals/festes-merce; Sep; free.*

🌐 Fêtes de Genève fireworks, Switzerland
No expense is spared when it comes to the fireworks display that marks the conclusion of Geneva's Fêtes de Genève. A fortnight of free concerts and cultural events climax with a spectacular hour of pyrotechnics above Lake Geneva. *fetes-de-geneve.ch; July-Aug; free.*

🌐 Fiesta de San Isidro, Spain
Annually, on 15 May, Madrid celebrates patron saint San Isidro with non-stop processions, parties and plenty of free concerts. Locals of all ages don their regional dress, and the night before is one of Madrid's biggest shindigs. *esmadrid.com; free.*

🌐 Hogmanay, Scotland
The trick to a low-cost Scottish New Year is avoiding the crowded (and expensive) ticketed events and finding a friendly pub, with a crowd of locals determined to get into the spirit of the occasion. *31 December; some events free.*

Below: yes, you do have to pay for beer at Oktoberfest in Munich, but you do get quite a lot of it.

EUROPE

International İstanbul Jazz Festival, Turkey

This offshoot of the İstanbul Festival is a major jazz (and friends) jamboree that has attracted top acts from Miles Davis and Dizzy Gillespie to Massive Attack, Björk and Lou Reed. Free events, like Jazz in the Parks, happen alongside ticketed ones. *caz.iksv.org/tr; Jul; some events free.*

King's Day, Netherlands

King's Day, on 27 April, is the bash of the year in Amsterdam, with boat parties, street celebrations, flea markets and lots of free activities and events. Wear orange in honour of birthday-boy King Willem-Alexander. *Iamsterdam.com; Apr; many events free.*

Las Fallas, Spain

Five days of fireworks, fire, food, fiestas and fun marks Valencia's Las Fallas, when each district of the city assembles its own surreal *falla* (float), before setting them alight in a grand conflagration to mark the arrival of Spring. *visitvalencia.com; Mar; free.*

Menningarnótt, Iceland

Every August, Reykjavík lets its flaxen hair down for Menningarnótt (Culture Night); museums throw open their doors, pop-up shops and bars erupt, and dance parties, concerts, performances and contests take place around town. *menningarnott. is; Aug; free.*

Notting Hill Carnival, England

London's biggest festival is an enthusiastic celebration of Caribbean culture, with outrageous costumes, riotously colourful floats and deafeningly loud sound systems. The streets of Notting Hill fill with the sound of popping Red Stripe cans and, ahem, herbal fragrances. *nhcarnival. org; free.*

Oktoberfest, Germany

The clue's in the name at this citywide celebration of beer and Germanic culture, which takes over Munich from September to early October. Many of the pop-up outdoor beer pavilions charge an entry fee, but bricks-and-mortar beer halls often don't, and there are free concerts for when you need a break from the *pils* and *helles*. *oktoberfest.de; Sep–Oct; some events free.*

Open House, Europe-wide

Want to peek into a London skyscraper? Cities across Europe – including London, Milan, Athens, Stockholm and Lisbon – hold annual Open House festivals, when access is granted to homes, offices and monuments that are normally closed to the public, for free. *openhouseworldwide. org; dates vary; free.*

Romarama festival, Italy

While many Romans leave the Italian capital in summer, folks who stay are rewarded with Romarama, a city-wide explosion of arts and culture. You'll find outdoor movies, musical acts, street art, concerts and much more. *romarama.it; Jun–Sep; many events free.*

Smaka på, Sweden

Snaffle bites from the Stockholm region's best kitchens (and vineyards, distillers and breweries) at this annual festival of food and drink, held in central Kungsträdgården park. *smakagoodfoodfestival.se; first week Jun; free.*

St Patrick's Day, Ireland

Forget the tacky St Patrick's Day celebrations outside Ireland; the real deal is *craic* (good times) incarnate. Dublin is one of the best spots to dive into the party, with parades, concerts (some free) and a lot of drinking. *Around 17 March; free.*

I'M DREAMING OF A (CHEAP) CHRISTMAS

Europe's Christmas and New Year celebrations are worth travelling for, particularly when sitting with a glass of warm mulled wine by a snowy-covered square, but prices can add up. To save, look for festive destinations that combine moderate room prices with inexpensive Christmas markets (Tallinn, Zagreb, Budapest, Prague), free Christmas church services (the famous service at King's College, Cambridge, is gratis if you queue early on the day; kings.cam.ac.uk) or cheap transport and fireworks you can watch for nothing (Berlin, London, Madeira).

EUROPE

OUTDOORS & ADVENTURE

Europe is mountains high and valleys low, fjords and forests, dunes and sea cliffs, silent lakes and urban beaches. The variety of landscapes – and variety of ways to enjoy them – has to be seen to be believed, and best of all, getting there is often the only expense.

NATIONAL PARKS & NATURE RESERVES

National parks, forest and nature reserves across Europe protect wild animals, threatened flora and epic landscapes, offering prime terrain for adventure. You'll

Above: cycle-touring around Europe is easier than ever with the Eurovelo bike routes. **Right:** taking surfing lessons in Portugal.

be pleased to hear that most are free to visit, though there are often fees for parking and camping. Read about them all at nationalparksofeurope.com.

• **Walk or pedal** – motorists will look on in envy as they pay for parking while you cruise by for free, on trails that lift you to within touching distance of the peaks.

• **Cool camping** – wild camping is only allowed in a few national parks, but there are inexpensive campgrounds run by park authorities in some truly spectacular locations.

• **Free digs** – many national parks and mountain areas provide free or cut-price accommodation for hikers, from refuges and *gîtes d'étape* in France (from €10; gites-refuges.com) and *rifugi* and *bivacchi* in Italy (rifugi-bivacchi.com; free or almost free) to Finland's vast network of free huts (nationalparks.fi).

• **Fab forests** – national parks get the fame, but national forests (managed by local government forestry departments) offer similar thrills with less competition

for space – and lower prices for accommodation and activities.

• **Carry the kit** – wildlife reserves across Europe have free hides for spotters with their own binoculars; bring a sleeping bag and stove and pay less for meals in mountain huts and refuges.

ADVENTURE ACTIVITIES

What Europe lacks in coral reefs and jungles, it makes up for in beaches, forests and mountains. National parks are the focus of attention, but many mountain areas are accessible from city centres for the cost of a ski-lift – meaning inexpensive starts to treks in summer.

• **Come off-season** – many mountain zones are expensive ski resorts in winter, but economical hiking bases in summer.

• **Choose the right side of the mountain** – ski in Tyrol and pay peak prices for passes, rooms and ski hire; cross into Italian Trentino and prices tumble; go east to Slovenia and they plummet.

• **Pick your pass** – prices for ski-lift passes vary wildly: in Italy's Abetone (abetone. com) you'll pay €25 per day; in Alta Badia (altabadia.org), you'll pay €58. Ski midweek to further cut your lift costs. Compare pass costs at onthesnow.co.uk/ europe/skipass.html.

• **Go off-piste** – cross-country skiing cuts out the pricey lift pass; find endless kilometres of trails in Austria, Switzerland, Italy and Scandinavia.

• **Go over it** – bolted climbing routes and vie ferrate (cable-protected climbing trails) across the Alps and Pyrenees are free to use if you have the ropes, harnesses and head for heights.

• **Get on your bike** – Europe's alpine roads turn getting to the next town into a stage on the Tour de France; trains across Europe carry bikes for free (or a modest charge) for the ascents you can't conquer.

HIGHLIGHTS

ECONOMICAL SKIING AT KRANJSKA GORA, SLOVENIA

The Sava Dolinka Valley offers top-class skiing conditions in the Julian Alps, at rates that significantly undercut the famous resorts further west. You can click into skis just minutes from the town centre, or cross-country ski to your heart's content. *p149*

SURF THE ERICEIRA COAST, PORTUGAL

The sole dedicated surfing reserve in Europe, this free-to-enter natural waterpark offers some of the most consistent surf in Europe, and gear is available to rent at prices that won't leave you running for dry land. *p155*

EUROPE

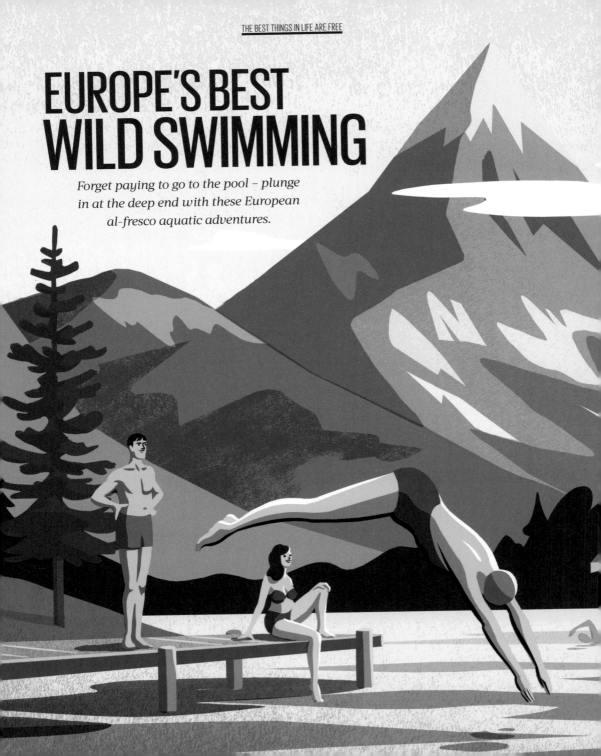

EUROPE'S BEST WILD SWIMMING

*Forget paying to go to the pool – plunge
in at the deep end with these European
al-fresco aquatic adventures.*

BAGGY POINT, ENGLAND

Swim through several secret sea caves at this delightful Devonshire dipping spot, including one tide-swept tunnel hidden from lily-livered landlubbers. Wildlife along this part of the coast includes seals, dolphins and porpoises. *nationaltrust. org.uk/baggy-point; Devon, England; free.*

UPPER LETTEN RIVERPOOL, SWITZERLAND

An urban oasis during the summer swelter, this 400m-long swimming canal in the fast-flowing River Limmat is a spot that attracts hundreds of swimmers. There's a 2m diving board, plus a bar and beach volleyball fields. *zuerich.com; Zürich, Switzerland; free.*

PURCARACCIA CANYON, FRANCE

The hike to the pools in this Corsican canyoning hotspot is demanding, but worth the effort. Swim in – and slide between – a series of pools fed by crystal-clear mountain water, and enjoy an infinity pool teetering on a waterfall edge. *Corsica, France; free.*

LAKE BOHINJ, SLOVENIA

The Julian Alps offer top-class swimming, including the lovely (but costly) Lake Bled lido. However, travel down the road to Triglav National Park's Lake Bohinj and discover wilder scenery for free. The Sava Bohinjka River provides a chilly thrill for flow-swimming fans. *tnp.si; Triglav National Park, Slovenia; free.*

LAKE WALCHEN, GERMANY

Bavaria is blessed with a batch of brilliant bathing spots, and this deep alpine lake in the mountains south of Munich is one of the most spectacular. Gravel beaches surround the mineral-rich turquoise water, and there are offshore islands to explore. *Near Kochel, Germany; free.*

LAKE ANNECY, FRANCE

Heated by subterranean hot springs, this alpine lake is unsurprisingly popular. Many lake beaches charge during the summer, but Plage d'Albigny and Plage des Marquisats are free. Wilder spots can be found too – some with great cliff jumps. *lac-annecy. com; Haute-Savoie, France; free.*

LAKE FIASTRA, ITALY

Camp beneath mountains on the north shore of this stunning beach-lined lake, and slide into the embrace of its gin-clear water. Take a snorkel – there's much to explore, including a rumoured submerged village. *lagodifiastra.it; Monti Sibillini National Park, Italy; free.*

LOCH LOMOND, SCOTLAND

You need a brave heart to swim in Scotland, but this beautiful loch – Britain's biggest – might tempt you in. It's the home of the annual Great Scottish Swim (and also, apparently, a monster, albeit one without the celebrity status of its Loch Ness cousin). *lochlomond-trossachs.org; Loch Lomond & the Trossachs National Park, Scotland; free.*

© Illustration | Owen Gatley

ALPINE ITALY & SLOVENIA

Forget Austria, Switzerland and France, the budget game in the Alps is in Italy and Slovenia. You get the same mountains and the same snow, but less hype and more bargains – for ski passes, accommodation and both summer and winter adventures.

✋ See Ibex at Gran Paradiso, Italy

You will indeed taste a little bit of paradise on the slopes of 4061m-high Gran Paradiso, where ibex prance in a stunning natural reserve that was once a hunting ground for King Victor Emmanuel II. Today, the hunters have been replaced by trekkers, skiers and climbers; many glorious, free-to-hike trails start at Cogne, and skiing costs from €8/27 (cross-country/downhill). *pngp.it; free.*

✋ Hike and bobsleigh up on Monte Mottarone, Italy

The cable-car trip up Monte Mottarone (1492m) offers eagle-eye views over Lake Maggiore and the islands of Bella and Superiore. Continue to the summit, where Alpyland has a 1200m-long bobsled descent with adjustable speeds for nervous riders. *alpyland.com/en; Verbano-Cusio-Ossola; cable car €19; bobsled €6 for one ride.*

✋ Cycling the Alpe-Adria, Italy and around

This transnational cycle route commences in Salzburg, Austria, and wends its way across some stunning countryside to the island of Grado off Friuli's coast in Italy. The Friuli section passes through Tarvisio, Venzone, Udine, Aquileia and Grado. In summer, special trains and coaches allow cyclists to hop on and off the trail. *alpe-adria-radweg. com; free.*

✋ Trek the Tre Cime Loop, Italy

The lonely, exposed sentinels known as the Tre Cime (Three Peaks) of Lavaredo are the classic image of the Dolomites. A not-too-gruelling 10km hike loops around the massif, offering uplifting views of the dolomite outcrops. Start at Rifugio Auronzo and circle round via Rifugio Locatelli (aka Dreizinnenhütte), where you can grab a beer and bite to eat in front of the view. *Free.*

Below: marvel at the agility of the ibex in Gran Paradiso natural reserve in northern Italy.

EUROPE

✋ Pedal the Fahrradroute Pustertal, Italy

More two-wheel fun is on offer in the Dolomites, where the mostly paved Fahrradroute Pustertal bike path follows Val Pusteria for 66km. Train stations and cheap bike rental locations at regular intervals along the way allow you to cycle segments and ditch the bike whenever you like. *pustertal.org/en/leisure-activities/ mountain-biking-and-cycling; bike hire per day from €18.*

✋ Driving the Stelvio Pass, Italy

The petrol isn't free, but the thrills come gratis on one of Europe's most epic drives, linking the towns of Stilfs and Bormio in northern Italy. The twisting, turning highway is the highest paved road in the eastern Alps, climbing to 2757m and offering predictably awe-inspiring views along the way. *Free.*

✋ Ride up, cycle down at Trento, Italy

Cable-cars swoop over the snows all over the Alps, but they often cost a pretty penny. Trento's Funivia Trento–Sardagna whisks passengers (and bicycles) up from downtown to the slopes above the river, where you can cycle, gravity-assisted, on winding roads back to town. *trentinotrasporti.it/viaggia-con-noi/funivia; cable-car fare €3.*

✋ Hike Triglav National Park, Slovenia

The 2864m limestone peak called Triglav (Three Heads) has been a source of inspiration and an object of devotion for Slovenians for more than a millennium. The area around Kranjska Gora and into Triglav National Park (tnp.si) is excellent for hikes and walks, ranging from the very easy to the difficult push for the summit. *Free.*

✋ Taking to the slopes in Kranjska Gora, Slovenia

Nestling in the Sava Dolinka Valley some 40km northwest of Bled, the Slovenian resort of Kranjska Gora is markedly cheaper than resorts in neighbouring Alpine nation. The main ski area is just five minutes' walk from the town centre, climbing the slopes of Vitranc mountain as far as Podkoren and Planica. There is also 40km of lovely cross-country trails. *kranjska-gora.si; day pass adult €34.*

✋ Hike the Soca River, Slovenia

Sliding through its rocky gorge, the Soca River is a winding line of perfect turquoise, and hiking trails run along its banks for 25km between Bovec and Trenta. It's one of Slovenia's most serene landscapes, and a fine back-route to reach Triglav National Park. *soca-valley.com; free.*

✋ Hire a rowboat on Lake Bled, Slovenia

Lake Bled may be Slovenia's most touristy spot, but there are inexpensive ways to enjoy it. A rented rowboat or stand-up paddleboard from the Castle Lido or campground will open up quieter corners away from the *pletna* tourist boats. And if you feel inclined, there's no charge to enter Bled Island (blejskiotok.si). *Rowboat hire €20 per hour, paddleboard hire €10 per hour.*

✋ Swim in Lake Bohinj, Slovenia

Just 26km from Lake Bled, but a very different experience, Lake Bohinj's chilly waters warm to a swimmable 22°C (72°F) in July and August. Swimming is not restricted and you can enter the water from any point on shore, though there are decent small beaches on both the northern and southern shores. Other gratis activities include hiking and cycling on lakeshore trails. *Free.*

TRAVEL THE IRON ROADS

Throughout the Alps, vie ferrate – literally, iron roads – traverse the mountains, a hangover from WWI, when messengers had to sneak across the peaks on foot to avoid enemy soldiers. Today, these cable-protected scrambling and climbing routes offer one of Europe's great adventures, with all the exposure of mountaineering minus the risk. Many are free if you have the gear (a climbing harness, a specialist lanyard, good boots and a helmet), including some heart-stopping routes over the Dolomites (see vieferrate.it for more).

EUROPE

BERGEN & FJORDS

With seven hills and seven fjords right on its doorstep, Bergen is the right kind of place for natural adventures. From the Unesco-listed weatherboard houses of downtown, you're just minutes from free swimming lakes, fjord-side beaches and more.

⚫ Float on a fjord pool at Nordnes
Beloved by locals of all ages, the Nordnes Seawater Sjøbad heated public pool is located in Nordnes Park, right on Bergen's harbour. It's a beautifully scenic spot to splash, catch some sun and have a picnic. *nordnessjobad.no; Nordnesparken; entry 80kr.*

⚫ Traverse Mt Fløyen overlooking the fjords
For an unbeatable Bergen view, ride the 26-degree Fløibanen funicular to the top of Mt Fløyen (320m). From the top, well-marked hiking trails lead into the pristine forest; grab a free walking map from the Bergen tourist office. *floyen. no; Vetrlidsalmenning 21; funicular return 95kr.*

⚫ Stay in a cabin in the woods
The Tubakuba forest cabin was built by architecture students using experimental wood-moulding techniques. Designed to help families with young children get back into nature, the cabin has no water or electricity but it's absolutely free of charge to stay here a night. Book through the Bergen Kommune office. *bergen.kommune.no; free.*

⚫ Swim at a Bergen beach
Bergen is ringed by idyllic swimming spots and beaches, easily reachable by bus. Try the rocky coves in the arboretum grounds at Milde, the beach behind the Norwegian royal family's residence at Gamlehaugen, or sandy, family-friendly Kyrkjetangen. *en.visitbergen.com/things-to-do/ attractions/beaches; free.*

⚫ Camp under the stars
About 20km from Bergen, between Espeland and Haukeland, the lovely lakeside campsite known as Lone Camping is a wonderful spot, far from the urban crowds. Bus 900 runs to/ from town for 53kr, and camping rates are reasonable too. *lonecamping. no; Hardangerveien 697, Haukeland; campsites from 165kr per tent, plus 15kr per person.*

⚫ Wander old Bryggen
Bryggen, Bergen's enchanting, multicoloured old quarter, runs along the eastern shore of Vågen Harbour, creating a free open-air museum of traditional Norwegian architecture. Each of the (mostly 18th-century) houses has stacked-stone or wooden foundations and rough-plank walls. It's most atmospheric when snow lines the streets in winter. *Free.*

Left: Bergen is the adventure gateway to western Norway.

© Grisha Bruev / Shutterstock

EUROPE

EUROPE'S
BEST PLACES
TO ESCAPE A CROWD

Yours to keep: a free pass to Europe's best bits – areas enshrined as national parks, where the only entry requirement is a sense of adventure.

OULANGAN KANSALLISPUIS-TO, FINLAND

Sitting on the Finland/Russia border, this stunning park is home to hundreds of reindeer and the multi-day Karhunkierros Trail, an 80-km trek during which hikers can stay for free in huts along the way. *luontoon.fi/oulanka; Lapland, free.*

OLYMPUS, GREECE

In Greek mythology the peaks and gorges of Olympus housed the 12 Olympian gods. Nowadays, golden eagles look down from the heavens, while the hills are home to wolves, wildcats – and walkers exerting themselves in a stunning mountain amphitheatre. *olympusfd.gr; Greece; free.*

GRAN PARADISO, ITALY

On the flanks of Italy's highest peak, 4061m-high Gran Paradiso, alpine ibex prance about like they own this national park. Once the hunting ground of King Victor Emmanuel II, this stunning park is now the preserve of hikers, climbers and cross-country skiers. *pngp.it; Italy; free.*

berto Moiola | Sysaworld/Getty Images

JOTUNHEIMEN, NORWAY

'Home of the Giants', this park contains the country's 29 highest mountains, including 2469m-high Galdhøpiggen. In Norway everyone enjoys *allemanns-retten* – free access to the countryside, including national parks. Backcountry hiking routes here include the Besseggen Ridge. *jotunheimen.com; Norway; free.*

BAVARIAN FOREST/ŠUMAVA, GERMANY/CZECHIA

This wild utopia is where Bavaria meets Bohemia in Europe's largest forest. Over 10,000 animal species inhabit these woods, which also have hiking, cross-country skiing and cycling trails. Wild camping allowed. *nationalpark-bayerischer-wald.de or npsumava.cz; Germany/Czechia; free.*

PARC NATIONAL DES PYRÉNÉES/PARQUE NACIONAL ORDESA Y MONTE PERDIDO, FRANCE/SPAIN

Separated only by the Pyrenees, these twin parks share fantastic fauna, including Europe's largest raptor, the lammergeier (bearded vulture). Hundreds of hiking trails wend through the valleys, and wild camping is possible on the French side. *pyrenees-parcnational.fr or ordesa.net; France/Spain; free.*

VATNAJÖKULL, ICELAND

Europe's biggest national park covers 12 percent of Iceland's entire surface area and is home to the continent's most powerful waterfall – the mighty Dettifoss. Visitors come to go glacier-hiking and ice-caving on the country's largest ice cap. *vatnajokulsth jodgardur.is; Iceland; free.*

BRECON BEACONS, WALES

This dark-sky reserve is home to the moody Black Mountains and magical Coed-y-Rhaeadr (Wood of the Water), where cascades include Sgwd-yr-Eira (the Snow Waterfall) on the River Hepste, with a path leading right behind the curtain of water. *breconbeacons.org; Wales; free.*

CAIRNGORMS, SCOTLAND

Britain's largest national park occupies a vast plateau punctuated by five of the country's highest peaks, including 1309m-high Ben Macdui, haunt of Am Fear Liath Mòr (a spectral beast) and several bothies (free mountain shelters). Not for the faint-hearted. *cairngorms.co.uk; Scotland; free.*

EDINBURGH & THE BORDERS

The Scots are fierce advocates for the great outdoors, and all manner of glen and loch adventures start within easy striking distance of the Scottish capital. You can even start the walk into the hills from downtown Edinburgh, putting the wilds within easy reach.

Ascend the sister vistas

For epic Edinburgh views, make the one-hour hike up Arthur's Seat, the hillfort-topped volcanic peak that dominates Edinburgh's eastern skyline. For a gentler jaunt, scale 143 steps to the brow of Calton Hill, where an unfinished Parthenon-style memorial overlooks the self-styled Athens of the North. *Free.*

Boulder at Salisbury Crags

The former hunting ground of Scottish monarchs, this green lung covers sheer crags, moorland and three pretty lochs. Members of climbing associations and clubs are permitted to climb on the South Quarry area of the Salisbury Crags, with some muscley bouldering routes and short ascents within sight of the city centre. *Queen's Dr; Free.*

Swimming al fresco

Swimming in Scotland? In summer, absolutely! Families gather to build mini-Edinburgh Castles on sandy Portobello Beach, while swimmers dive into the chilly waters of the Firth of Forth. True bravehearts can join the Loony Dook (free), a New Year's Day dip in the Firth at South Queensferry.

Nature walks on Blackford Hill

The wooded Hermitage of Braid and Blackford Hill are crisscrossed with walking trails – with sunlight filtering through the leaves and birdsong all around, you'll feel miles from the city. Check routes at the visitor centre in Hermitage House. *walkhighlands.co.uk/lothian/blackford-hill.shtml; free.*

Go underground

While ghost tours of Edinburgh's underground vaults have become mainstream, Gilmerton Cove remains an off-the-beaten-track secret. Hidden in the southern suburbs, the mysterious 'cove' is a series of subterranean tunnels hacked out of the rock for purposes unknown. *gilmertoncove.co.uk; 16 Drum St; tour £7.50.*

Hike vast landscapes in the Pentland Hills

Rising like a wave on the southern edge of Edinburgh, the rugged Pentland Hills exert a strong magnetic pull for walkers. Reaching 579m, the high ground offers excellent – but not knee-wobbling – walking with stunning views. Buses run from Edinburgh Bus Station to Biggar in the heart of the hills. *pentlandhills. org; free.*

LISBON & THE ATLANTIC COAST

Hilly Lisbon already hints of the outdoors adventures just beyond the city limits: surfing on blissful beaches, hiking to urban viewpoints and more. Inexpensive transport links mean it's easy to get back for a bedtime pastéis de nata (custard tart) at the day's end.

Surf Ericeira

Picturesquely draped across sandstone cliffs, sunny, whitewashed Ericeira is a playground for surfers, who come here for the great waves and camaraderie. Declared in 2011, it's the sole dedicated surfing reserve in Europe: the Activity Surf Centre rents out boards and wetsuits, and morphs into a bar at night. *activitysurfcenter.com; Rua de Santo António 3A; per day surfboard/wetsuit hire €20/10.*

Skate alongside the surf

Operated by Lisbon's flagship Quiksilver store, the skate park right alongside the beach at Ericeira is free to use, so you can ride the ramps, rails and bowl for as long as you like; be prepared: the level of local talent is pretty high. Skate competitions take place here regularly, or there are yoga classes nearby on the lawn. *quiksilver.pt; Av São Sebastião 36A; free.*

Hit the beaches

The sandy arcs west of Lisbon, on the Tagus estuary's northern bank, are the city's most accessible beaches, with gentle surf, soft sand and good train services. Try Praia de Carcavelos and Praia do Tamariz for people-watching and castle-building, or Praia do Guincho and Costa da Caparica for surfing. *Free.*

Take in a view

The city of seven hills serves up a septuplet of *miradouros* (viewpoints). The highest is Miradouro da Senhora do Monte, Our Lady of the Hill, looking out over Mouraria Moorish Quarter and the Tagus estuary. More free vistas are served up at the Miradouro de São Pedro de Alcântara, reached via the Ascensor da Glória funicular. *Free.*

Explore Sintra-Cascais Natural Park

The westernmost point of mainland Europe, Cabo da Roca ends at cliffs offering one of Portugal's best free views. Further north, preserved dinosaur tracks wander up the chalky south cliff behind Praia Grande, left by megalosaurus and iguanodons 100 million years ago. *Free.*

Below: Ericeira is a venue for pro surfing competitions, but beginners can also rent equipment and pick a smaller wave to start out on.

EUROPE

© homydesign | Shutterstock

EUROPE'S BEST FREE WALKING TOURS

Yes, that's right, a guided city stroll, for free (apart from the discretionary tip at the end).

ALTERNATIVE BERLIN, GERMANY

See all the great, gritty bits of the German capital: street art, skateparks, artist squats, multicultural neighbourhoods, graffiti galleries, day raves, flea markets and more. *alternativeberlin.com; daily.*

RUNNER BEAN TOURS, SPAIN

Architect Gaudí made a big mark on Barcelona; this freebie walk focuses solely on him, from his first masterpiece, Palau Güell, to his unfinished magnum opus, the Sagrada Família. *runnerbean tours.com; daily.*

NENO & FRIENDS, BOSNIA & HERCEGOVINA

Let Neno and Merima show you the must-see attractions of Sarajevo, offering a personal and passionate take on the country's history, politics and thriving cultural scene. *sarajevowalkingtours. com; time varies, Apr–Oct.*

BRUSSELS GREETERS, BELGIUM

This goes one step further than a regular gratis city tour, as you meet up with these volunteers in advance to arrange a bespoke freebie, based on your interests – history, football or *frites*. Tips not required. *visit.brussels/ en/sites/greeters*

NEW ROME FREE TOUR, ITALY

What we like about this is its timing. The afternoon tour leaves the Spanish Steps for a scoot around key sights (Temple of Hadrian, Pantheon, Trevi Fountain); good for solos who'd like evening company. *newromefreetour. com; times vary.*

BELGRADE WALKING TOURS, SERBIA

Start with the Downtown Tour for an intro to the Serbian capital, then join the Zemun Tour (Saturdays), which focuses on the narrow streets of this old neighbourhood. *belgradewalking tours.com; daily.*

WALKATIVE TOURS, KRAKÓW, POLAND

Choose the classic Old Town Tour; moving Jewish Kraków; pickle- and *pierogi*-filled Foods of Kraków; or late-night Macabre Kraków. *freewalk ingtour.com/krakow; hours vary.*

LJUBLJANA FREE TOUR, SLOVENIA

Meet a yellow-dressed guide at the Pink Church for a tour of the Slovenian capital, both its main sights and the secret bits only a local knows. *ljubljanafree tour.com; daily.*

FREE WALK ZÜRICH, SWITZERLAND

If you think lakeside Zürich is all cold corporate finance, join the Zürich West tour, which lifts the lid on the city's grungier side – from factories-turned-theatres to the red-light district. *freewalk.ch/ zurich; Fri, Sat & Sun.*

BATH GUIDES, ENGLAND

The Mayor of Bath's Honorary Guides have been running free tours of the Georgian city for 90 years. Tours start outside the Pump Rooms. *bathguides. org.uk; daily.*

WELLNESS

Europe's wellness traditions offer a naturalistic take on whole-body health. Natural springs form the bedrock (or should that be form in the bedrock?) of European wellness, from boil-an-egg-hot fumaroles in volcanic Iceland to sculpted calcite landscapes at Pamukkale in Turkey and cavern baths at Miskolctapolca in Hungary.

Then there's getting hot and sticky: Europe's tradition of steam baths, hammams and saunas dates back to at least ancient Greek times, if not earlier. If you stick to civic bathhouses, guesthouse

Above: public bathing in Budapest. Right: or find more private wild swimming spots in Finland.

saunas and natural hot springs, you'll never have to pay spa prices to get warm, detoxed and well.

HOT & COLD SPRINGS

Plate tectonics have created hotspots all over Europe, from the fire-and-ice natural springs of Iceland to the theatrical public baths of Budapest. Budget travellers take heart: for every swish spa, there's a public spring or pool where you can bathe in the same water for free (or at least for modest fees).

• **Carry a towel in volcanically active areas** – you never know where you'll spot steam billowing, marking a natural hotspot (just check the water temperature first!). In Iceland, trade the pricey Blue Lagoon for the free public thermal pools along the southern side of the island in Reykjavík, Landmannalaugar, Reykjadalur and Seljavallalaug.

• **Hungary scores highly for thermal springs** – most rush to Budapest, but don't overlook Hévíz (heviz.hu) where modest fees get you access to Gyógy-tó, Europe's largest thermal lake.

• **Go off the beaten track** – some of Europe's best and cheapest thermal

springs are off the tourist trail; escape the tourist crowds at Llixhat e Bënjës in Albania or Băile Herculane in Romania.

• **Don't overlook ancient Greek and Roman springs** – such as Bath in England, Pamukkale in Turkey and the free-to-use Terme di Saturnia in Tuscany, still busy after 2000 years.

NORDIC TRADITIONS

Wellness flows in the veins of Nordic countries, particularly after a stimulating sauna or ice-swim. From natural, volcanic hot springs to public swimming pools heated by geothermal boreholes, there's hot water everywhere, and winter brings the added thrill of leaping from your sauna and diving through a hole in the ice. The best bit? Getting hot, cold or wet (or all three) is often gratis.

• **Choose your sauna** – in many places, traditional smoke saunas (where smoke and steam fill the room) are more expensive than smoke-free electric saunas.

• **Steam at your hotel** – there's no need to visit a spa; most hotels in countries such as Finland have an on-site sauna that guests can use for free.

• **Look to the lakes** – free lake swimming is everywhere, even in the centre of Scandinavia's biggest cities. Some lakes even come with public saunas at the water's edge.

• **Get spiritual** – some of Scandinavia's natural experiences are almost religious – watching the aurora borealis, for example, or combing Iceland's undergrowth for elves (they exist, locals insist).

• **Forage** – the endless forests of Scandinavia unleash a health-giving bounty from July to October in the form of blueberries, crowberries, raspberries and cloudberries.

HIGHLIGHTS

HOT RIVER SWIMMING AT REYKJADALUR
You have to hike through hot sulphurous plains to reach the hot river at Reykjadalur, but on arrival you can lower yourself into the balmy waters, even when snow is falling. The flow of the water carrying your cares away adds some extra magic. *p160*

BREAKING THE ICE AT KUUSIJÄRVI
Ice-swimming has a long history in Finland, and at this scenic lake near Helsinki, you can combine getting extremely cold with getting extremely hot – both activities are said to stimulate the circulation and invigorate the body. Do you feel brave? *p163*

EUROPE

REYKJAVÍK & SOUTHERN ICELAND

The Icelandic capital is expensive, but bucket-list experiences such as swimming in volcanic hot springs and watching the northern lights don't need an expense account. Save your krona for food and accommodation, and enjoy a free serving of natural wellness.

📍 Nauthólsvík Geothermal Beach

Tucked behind the university and Reykjavík's mini-airport, this small sandy stretch edges into the chilly Atlantic, but geothermal water is rerouted to keep the lagoon between 15°C and 19°C. As an added perk, a public hot-pot simmers at 38°C. There's a fee in winter, but summertime is free. *nautholsvik. is; Nauthólsvík; closed Fri & Sun; free in summer.*

📍 Watching the aurora borealis

Nobody has found a way to charge for the eye-popping light show that Iceland puts on every winter, courtesy of the northern lights. Watching this incredible, pulsating visual display is a profoundly spiritual experience that will leave you considering your place in the universe. See it from Þingvellir National Park, a great spot for dark skies. *Rte 36/Þingvallavegur; Sep-Mar; free.*

📍 Hunt for elves in Hafnarfjörður

Amongst Icelanders, Hafnarfjörður is famous for its population of elves and dwarves – seriously. The areas frequented by the hidden folk are said to be connected by spiritual lines of power, including the Hellisgerði Lava Park (Hellisgata 3; free), where you can seek the assistance of elf experts on guided tours (alfar.is; Tue & Fri summer; 4500kr).

📍 Bask in a hot river at Reykjadalur

A 3km walk through sulphur-belching fields leads to the delightful geothermal valley of Reykjadalur, famed for its 'hot river' where bathers can strip off and soak in naturally hot mineral waters (bring swimwear). The tourist information office in Hveragerði can help you find the trail. *Breiðamörk; free.*

📍 Have a warm swim in stunning nature at Seljavallalaug

Keflavík's Blue Lagoon is a pricey day out, but you can enjoy the spa pool experience gratis at one of Iceland's oldest geothermal bathing pools. Set into a rocky gully beside a river, this is a serene spot in which to acquire a warm glow of wellbeing in 20-30°C water. *Seljavallavegur, off Rte 242; free.*

Below: an urban geothermal beach in Reykjavík, warm even in winter.

© Try_my_best | Shutterstock

EUROPE

STOCKHOLM & THE ARCHIPELAGO

Stretching across 14 islands, Stockholm sheds its smart threads when it comes to wellness. Locals are happy to periodically trade the cool furniture for freebies from nature: lake swimming, wild foraging and nature-reserve saunas, all accessible from downtown.

🖐 Go foraging in the woods

With its forested hillocks, dreamy Djurgården island is a short stroll from downtown. Its calm woods make an ideal setting for a picnic, a restorative stroll and some late summer foraging for blueberries, cloudberries and chanterelle mushrooms. Just make sure you know what you're picking. *Djurgården; free.*

🖐 Climb to cliff top views

A tonic for the city sprawl, the views from cottage-lined Monteliusvägen are soul-affirming – particularly in winter, when a crazy-paving of broken ice fills Riddarfjärden lake. Follow the rickety wooden handrails that skirt round the hillside just north of Bastugatan, due west of Slussen station. Now sit back, look over the church spires, and exhale. *Slussen T-bana station; free.*

🖐 Slow down with island life

It's a gentle 25-minute boat ride to Fjäderholmarna (the 'Feather Islands'), where locals come to slow the pace of life. Swim off the rocks in clean, bracing waters, then stuff yourself on smoked fish at waterside restaurants. Go on, unleash your inner Scandinavian! The ferry leaves from Nybroplan. *visitstockholm.com/ stockholm-archipelago/fjaderholmarna; May–Sep; entry free, ferry 145Kr.*

🖐 Join in with free alfresco aerobics

In the warmer months, Rålambshovsparken is packed with picnicking Swedes fresh from a dip in Riddarfjärden lake. Take a swim, hire a canoe or get physical at the free outdoor aerobics sessions – there's nearly always a free nightly class of some kind during the summer. *Free.*

🖐 Take a lakeside sauna

After a healthy hike around forested Källtorpssjön lake, a 15-minute bus ride from Slussen station, sweat it out with a lakeside sauna at Hellasgården, perhaps with a quick burst in the frozen lake in winter. There are separate male and female saunas at a cheap hourly rate; expect a nude dress code. *hellasgarden.se; sauna from 70-100kr.*

Below: enjoying a picnic in Stockholm's Djurgården, easily reached by bicycle from the city.

EUROPE

BUDAPEST

Nowhere are spa traditions more grandly executed than in Budapest, a capital of two cities bisected by the blue Danube. You can pay peak rates for the most lavish bathhouses, or get almost the same grandeur on a budget at smaller, less flashy spas.

Soak under the sky

The Széchenyi Baths are the largest medicinal baths in Europe, and there's a lot included in the not-too-onerous admission fee: 15 indoor thermal pools with water temperatures up to 40°C, and three outdoor pools, including a whirlpool. Early mornings and weekdays see cheaper rates. *szechenyispabaths.com; XIV Állatkerti körút 9-11; weekday rates 5600-6800Ft.*

Amble by a river on Margaret Island

Smack bang in the middle of the Danube, largely pedestrianised Margaret Island is a 2.5km-long retreat offering peaceful respite for locals and tourists on hot summer days. The swimming pools, running track, bicycle hire and bars require you to flash the forint, but it doesn't cost a thing to walk or jog and soak up the fresh air in the gardens. *Free.*

Soak in grand surrounds at Gellért Baths

A dip in these Art Nouveau baths has been likened to taking a swim in a cathedral. The indoor baths, open year-round, are the most beautiful in Budapest, and from May to September you can splash in the outdoor wave-pool in waters heated to 26°C, surrounded by landscaped gardens. *gellertbath.hu; Kelenhegyi út 4, 1118; from 5600Ft.*

Catch the sun from Gellért Hill

Come here for therapeutic views of Buda, Pest and the Danube, best at sunrise or sunset. The hill is topped by the Citadella fortress, built by the Habsburgs after the War of Independence (1848–49) but never used in battle; and the 14m-high Liberty Monument (1947), honouring the Soviet soldiers who liberated the city in 1945. *Free.*

Splash and drink at Lukács Baths

Housed in a 19th-century complex, Lukács offers baths, saunas and outdoor pools at reasonable prices. The thermal pools simmer at 22°C to 40°C, and on weekdays you can also take a drinking cure at the entrance. *lukacsbath.com; Frankel Leó út 25-29, 1023; sauna 600Ft, baths 4300Ft.*

Below: the ornate interior of the Széchenyi Baths is worth the price of admission.

© posztos | Shutterstock

HELSINKI

*Brand new by European capital standards, 200-year-old Helsinki's walkable downtown
is easy to enjoy on the cheap. In summer, Finns pack the seaside beaches, islands and
cafes, retreat for a revitalising sauna, then let loose till late under the midnight sun.*

Take a sauna

The most Finnish thing you can do
in Helsinki is ubiquitous and requires
nothing. Not even clothing. Almost every
building in the country has a sauna of
some kind, and that includes hotels, inns
and hostels, where steaming is often free
to guests in temperatures that average
70–90°C. *Some saunas charge.*

Breathe in the sea air on Suomenlinna

Reached by ferry from the quay at
Kauppatori, this fortress island is studded
with heritage architecture and military
history, but locals love it best as a place to
picnic, sunbathe and splash in the clean
waters of the bay – the perfect tonic to
city life. There's even a hostel for cheap
overnight stays. *suomenlinna.fi; Port of
Helsinki; ferry ticket return €5.*

Bask under the midnight sun

During the endless nights of summer,
locals take advantage of the late, late
nights to talk, talk, talk. A long midsummer
session of chat and liquorice-flavoured
vodka can be highly therapeutic –
consider heading to Ravintola Skiffer on
Liuskaluoto island (skiffer.fi; Liuskaluoto), a
restaurant on a former fuel station in the
bay, with good swimming nearby and its
own sauna.

Ice swimming at Kuusijärvi

There are two types of people: those who
think nothing of diving into a hole in the ice
in mid-winter, and those who wouldn't do
it for all the Koskenkorva (vodka) in Finland.
Lovely Kuusijärvi lake, just outside Helsinki at
Vantaa, is a thrilling place to take the plunge,
with the option of a sauna. *cafekuusijarvi.fi;
swimming free, sauna from €6.50.*

Splash at Hietaranta

Helsinki has plenty of pretty beaches, but
the best is the golden, sandy stretch due
west of the city centre at Hietaranta – ideal
for fresh Baltic swimming or just paddling.
The most appealing way to get here is to
stroll west from Mechelininkatu through
Hietaniemi cemetery. *Töölö District, free.*

Below: by most
indices, Helsinki has
one of the world's
highest quality-of-
life ratings, not least
due to its natural
setting.

EUROPE

THE BEST THINGS IN LIFE IN NORTH AMERICA: TOOLKIT

The continent of North America – the US, Canada, Central America and the Caribbean – can be painfully expensive or gratifyingly good value, depending on how many top experiences you plan to clock up. Here's how enjoy big cities and backwaters alike.

North America is a place where it's easy to spend money and tricky to save, thanks to expensive accommodation and high entry fees to many sights, but the cost map of the continent is highly uneven. Big US cities such as New York and San Francisco rank amongst the most expensive travel destinations on the globe, while sleepy small towns in Central America may leave you grinning at the manageable cost of living.

TOP BUDGET DESTINATIONS IN NORTH AMERICA

There's budget travel gold in them hills if you know where to look – here's our top five:

• **WASHINGTON DC:** Maximum bang for your cultural buck in the US capital.
• **AUSTIN, TEXAS:** Friendly, foodie and fun to be around, with plenty of ways to save.
• **WASHINGTON STATE**: Free adventures by the truckload for travellers with rented wheels.
• **VANCOUVER:** The big city with big (and free) adventures on the doorstep.
• **MEXICO CITY:** Cheap street eats, high-octane culture and plenty to see for free.

TRANSPORT

Transport prices can vary widely across this enormous region. Central America is a thrifty traveller's paradise, with cheap buses (often retired US school buses) zipping everywhere, but prices raise as you head north to the US and Canada.

Getting anywhere means putting in the miles, whether flying, driving or taking the Greyhound. But there are still ways to save: take cheap flights, use the buses,

Below: with great food and amazingly accessible adventures, Vancouver is a top destination for outdoor types.

© Pashu Ta Studio | Shutterstock

offset car-hire costs by camping and skipping organised tours, and your money may go further than you think.

AIR

Discount airlines have proliferated in the US in recent years. The big budget players are Southwest Airlines (southwest.com), Allegiant Air (allegiantair.com), Frontier Airlines (flyfrontier.com) and JetBlue Airways (jetblue.com), with a handful of local discount carriers covering Canada and Mexico.

• **Play the market** – don't treat a plane ticket as a done deal; if the price for the route drops, you may be able to cancel the ticket without a huge penalty and rebook at the lower fare.

• **Get a pass** – all the big airline groups have passes that offer discounts on local hops when booked with an international flight, including SkyTeams's Go America & Canada Pass (skyteam.com), Star Alliance's North America Airpass (staralliance.com) and Oneworld's Visit North America Pass (oneworld.com).

TRAIN

Train routes fan out across the US and Canada but fizzle out past the Mexican border. Amtrak (amtrak.com) has routes across America, but its trains are rarely the quickest or cheapest way to explore.

• **Use your ID** – students and seniors over 62 get 15% off Amtrak fares. Canada's trains are operated by VIA Rail Canada (viarail.ca), with similar deals.

• **Take commuter trains** – they're arguably more useful than long-distance trains, particularly on the northeast coast of Canada and the US. Avoid expensive, express Acela unless you have to hurry.

• **Time it right** – book long-distance trains on weekdays, commuter trains at weekends; and buy tickets at least a week

in advance for the lowest fares.

• **Pick up a pass** – Amtrak's USA Rail Pass offers discounted coach-class travel for 15 to 45 days, covering eight to 18 train journeys. Canada's Via Rail (viarail.ca/en/fares-and-packages/rail-passes) has a handy Canada Pass valid for 15 to 60 days.

BUS

Buses run cheaply all over North America, and tickets are a bargain in Central America. But there's a certain snobbishness about taking the bus in the US and Canada, where the private car is king.

• **Go by Greyhound** – the US' and Canada's most famous buses (greyhound.com) have toilets, free wi-fi and power outlets, at rates that undercut planes, trains and automobiles. Seniors over 62 years get 5% off; students get 10% off with a Student Advantage Discount Card (studentadvantage.com; US$23).

• **Try the competition** – local bus companies under the Trailways banner (trailways.com) cover some destinations Greyhound doesn't.

• **Bargain buses** – book well ahead for the best fares with discount companies like Megabus (megabus.com) and BoltBus (boltbus.com).

Above: Amtrak trains are an interesting alternative to taking a flight if you wish to see more of the country from your seat.

NORTH AMERICA

TRAVEL BY CAR

Don't kid yourself, you are not getting around Canada and the US without an automobile, and a rental car will come in handy in Central America too. Encouragingly, however, rental prices are low (from US$25) and petrol is cheaper than Coke (by volume). You'll burn through the stuff on any long hop, so park those dreams of driving a Cadillac and stick to a small, economical compact.

• **Things to skip** – return where you picked up to avoid extra charges and forgo airport pick-ups for cheaper downtown rentals. Weekly rates offer savings on daily rates, but only rent for days you actually need to drive: city parking costs can be brutal.

• **Be a driveaway driver** – several companies offer one-way car hire, relocating vehicles across the country for people who are moving home; check what's available at autodriveaway.com/driver.

• **Share the ride** – for trips by RV (recreational vehicle), look at peer-to-peer rental networks such as Outdoorsy (outdoorsy.com) and RVShare (rvshare.com).

LOCAL TRANSPORT

Most big cities have reliable urban transport, using trains (underground, elevated or at surface level) plus buses, taxis and sometimes boats. Prices are high for taxis, low for mass transit, and get cheaper if you buy a multi-journey pass or pre-payment card.

• **Taxi tips** – see hailing a cab in the street as something of a luxury; save in the US and Canada with rideshare apps such as

Above: the US (including Death Valley here) and Canada are the spiritual home of the road trip – renting a car (and fuelling it) is more affordable here than in many parts of the world.

Uber, Lyft and Gett, or Uber and Cabify in Central America.
- **Carpool** – for fixed routes, share a ride with others via a carpooling app such as Via, or the pooling services on Uber and Lyft.
- **Pedal power** – use inexpensive bike-share schemes such as New York's Citi Bike (citibikenyc.com) and Toronto's Bike Share (bikesharetoronto.com).

ACCOMMODATION

Your biggest daily cost in North America will be for the time you're asleep. Central America lays on plenty of budget beds, but elsewhere you can bank on paying US$50 or more for a room, even at the budget motel end of the market – and you can easily triple that in cities such as New York.

CAMPING

Camping is a great solution to the accommodation conundrum, but you can't just turn up anywhere and hammer in tent pegs. Even national parks restrict camping to paid campgrounds.

In the US, the local terminology for camping outside of designated sites is 'dispersed camping' (you'll also hear 'dry camping' and 'boondocking'). For cheap or free camping, look to national forests and national grasslands (fs.usda.gov) and land administered by the Bureau of Land Management (blm.gov) – the FreeRoam app (freeroam.app) can help you find locations. In Canada, look for 'Crown Land' in each of the country's provinces.
- **Camp the parks** – most national parks (find them all at nps.gov) have campgrounds (with spaces priced from US$15-30, up to US$50 for RVs). Make bookings well ahead at recreation.gov.
- **Get a pass** – the America the Beautiful Pass (store.usgs.gov/pass; US$80) covers a year of free entry to all of America's national parks, plus camping discounts.
- **Find a forest** – national forests are cheap (or free) alternatives to camping in national parks (try Kaibab National Forest if all the Grand Canyon National Park sites are full).

HOSTELS, HOTELS & GUESTHOUSES

Sleeping bargains exist in the Americas – particularly in Central America, where dorm beds can be had for loose-change prices – but don't rely on being able to find a wallet-sparing room at every stop. During the holidays – summer, Thanksgiving, Christmas and Spring Break – Americans move around their nation (and neighbouring Central America and the Caribbean) in incredible numbers and cheap rooms book up months in advance.

Hostels aren't as common as in Europe or Asia, but you'll find them dotted about in the US and Canada, and more frequently in Central America – many are members of Hostelling International USA (hiusa.org). Look for space at hostelbookers.com and hostelworld.com.
- **All hail the highway motel** – cheap chains cluster around Interstate highway exits and on the outskirts of towns and cities in the US and Canada; Comfort Inn (choicehotels.com), Red Roof Inn (redroof.com) and Motel 6 (motel6.com) offer bargain rates.
- **Budget B&Bs** – B&Bs undercut hotels, and throw in a home-cooked breakfast; for more informal rooms to rent, Airbnb (airbnb.com) has extensive listings.
- **Lodge in the hills** – lodges aren't cheap, but US$100 or so at one will get you a room in the heart of many national reserves; to save, stay just outside the parks and visit on day-trips.

NORTH AMERICA

MOVIE NIGHTS

EN PLEIN AIR, MONTRÉAL, CANADA

Feeling curious? Catch some of the world's best documentaries for free at screenings curated by the Montréal International Documentary Festival at parks and other locations. *ridm.ca; screenings throughout the week, Jul-Sep.*

CAPRI DRIVE-IN THEATER, COLDWATER, USA

A slice of Americana between Detroit and Chicago, operating since the 1960s, the twin-screen Capri has retained a cool vintage vibe. From the long car line-ups, to the sound system (your radio) it's worth the $10 ticket price. *capridrive-in.com; 119 West Chicago Rd; weekends, sunset.*

NEW ORLEANS FILM SOCIETY: MOONLIGHT MOVIES, NEW ORLEANS, USA

An exquisite curation of mostly free, outdoor movies at unique venues across the city (Old US Mint; Sculpture Garden) ramps up the cultural quotient. *neworleansfilmsociety.org; spring and fall schedules.*

MIAMI BEACH SOUNDSCAPE CINEMA SERIES, MIAMI, USA

Palm trees and bougainvillea provide the setting for weekly, free, family friendly movies on a huge projection wall at this urban park, steps from Lincoln Road. *mbculture.com; Soundscape ExoStage, 17 St and Washington Ave; Wed, Oct-May.*

OLD PASADENA FILM FESTIVAL, PASADENA, USA

Once home to Hollywood's earliest movie stars, Pasadena now offers a free month-long festival featuring an eclectic selection of movies at several venues. Locals grab snacks from the flocking food trucks. *oldpasadena.org; Jul-Aug.*

FREE FRIDAY FILMS AT INNIS TOWN HALL, TORONTO, CANADA

Well connected with Toronto's many film festivals, University of Toronto's Cinema Studies Students' Union delivers advance screenings, art films, and director visits. *cinssu.com/fff-links. Innis Town Hall, 2 Sussex Ave; Fri, Sep-Apr.*

MOVIES WITH A VIEW, BROOKLYN, USA

Brooklyn Bridge Park presents a mash-up of classic and cult films, against a suitably cinematic backdrop of the Manhattan skyline. Hipster meets Boho meets Woody Allen... only in New York. *brooklynbridgepark.org; Brooklyn; Thu at sunset, Jul-Aug.*

PHILADELPHIA FREE SCREENINGS, USA

Philadelphia lays on food and ice cream trucks to make an occasion of its free film screenings, held every summer as the sun sinks in parks, playgrounds, malls and cafes. Movies tend to be family friendly blockbusters, and everything but the snacks are free. *visitphilly.com; sunset, May-Sep; free.*

CINESPIA, LOS ANGELES, USA

The Hollywood Forever Cemetery (hollywoodforever.com; 6000 Santa Monica Blvd) is just one of the inspired locations for the summertime movies screened by Cinespia. There are always free options amidst the ticketed events. *cinespia.org; venues vary; May-Sep; some screenings free, $45 per car for drive-ins.*

FRESH AIR CINEMA, VANCOUVER, CANADA

Outdoor movies are screened for free at venues around Metro Vancouver every summer, including in Stanley Park, with thousands of blanket-hugging locals. Check ahead to see what's on during your visit. *freshaircinema.ca; free.*

© Illustration | Thomas Burden

ARTS & CULTURE

The United States has a huge presence on the world's cultural stage, thanks to the Hollywood movie machine and the American music industry. However, don't assume the US is the only show in town: Canada, Central America and the Caribbean have their own rich creative traditions, easily (and cheaply) experienced at museums, galleries and historic sites across the region.

See-it-before-you-die museums and monuments are as common as stop signs in many corners of North America, but seeing them for free is the challenge.

Above: Mexico City's world-class Museo Nacional de Antropología. Right: a view, here of Chicago, often comes with a cost in the US.

More sights have entry fees than don't, but there's plenty of free culture if you know where to look, from top-tier museums such as LA's Getty Centre to ancient Maya ruins in Mexico and Central America.

AMERICA'S TOP SWITCHEROOS

Prices are steep at top sights, but it's easy to trade bank-breaking attractions for less well-known sights and get the same wow for less bucks.

- **SKIP:** Statue of Liberty, New York, USA (entry US$18.50-21.50)
- **SEE:** Mount Rushmore, South Dakota, USA (entry free, parking US$10)
- **SKIP:** Empire State Building, New York, USA (entry US$38-58)
- **SEE:** Los Angeles City Hall Observation Deck, USA (entry free)
- **SKIP:** Los Angeles County Museum of Art, USA (entry US$25)
- **SEE:** Getty Center, Los Angeles, USA (entry free)
- **SKIP:** Chichén Itzá, Yucatán, Mexico (entry U$23.40)

- **SEE:** Uxmal, Yucatán, Mexico (entry US$3.75)
- **SKIP:** City views from the CN Tower, Toronto, Canada (entry C$38)
- **SEE:** City views from Terrasse Dufferin, Québec, Canada (free)

NORTH AMERICA'S CULTURE FOR LESS

The US is home to many of the world's most famous museums, galleries and monuments – the Statue of Liberty, the Kennedy Space Center, the Metropolitan Museum of Art – but don't overlook less celebrated marvels in Canada, Central America and the Caribbean: they're just as interesting and often free.

At the most prestigious addresses, culture comes at a price, and that price can be upwards of US$30 per person. For a more frugal dose of culture, even big-bucks locations such as New York and Los Angeles have free museums, or free opening days (many museums offer gratis entry one day per week or month). Alternatively, head north or south of the border; prices settle at more manageable levels in Canada and Central America.

- **Monuments versus National Monuments –** if there's a view, there's normally a fee, and this includes most of the US' famous skyscrapers. As an alternative, the views are gratis at many civic buildings and national monuments, from Mount Rushmore to the National Mall. See nationalparkobsessed.com/list-of-national-monuments for a full list.
- **Flash your ID –** students get bargain prices at many museums, galleries and monuments; the International Student Identity Card (ISIC; isic.org) is widely recognised.
- **Be forever young –** many museums are deliberately kept free for young

Amusement parks and fairgrounds are an integral part of American life, but a trip to the country's biggest and best-known is big-bucks entertainment – don't expect change from a Franklin bill (US$100) for a day at one of these. Look to smaller, regional parks instead, such as the Six Flags parks (sixflags.com) in the US, Canada and Mexico, which offer discounted entry if you collect coupons from promotional packs of drinks and snacks. Alternatively, swing by the state fair – most states have them, in late summer or fall, and there's always a fairground.

people (sometimes everyone up to 25), particularly in Canada.

• **Wonderful Washington** – the US capital has the greatest density of free museums in the country, if not the world, including the 19 museums and zoological park run by the Smithsonian Institute (si.edu).

• **Small-town wonders** – many of America's most interesting, eccentric museums are tucked away in small towns, with small entry fees – consider the Museum of the Potato (idahopotatomuseum.com) in Blackfoot, Idaho, or the Museum of Bad Art (museumofbadart.org) in Sommerville, Massachusetts.

• **Government buildings** – many of America's government institutions are free to visit, including the White House and Capitol in Washington, DC, though there are some administrative hoops to jump through.

AMERICANA

Nowhere has built a legend around itself quite like the USA, and sights such as Route 66, the Hollywood sign and the crossroads in Clarkesdale, Mississippi (where Robert Johnson allegedly sold his soul to the devil) are wonders as much for what they represent as for what there is to see.

Classic Americana is free (or cheap) to observe all over the country, from classic drive-ins (p169) to rural diners, Art Deco motels and wacky expressions of small town civic pride. Plot a road trip round these classic sights and you've got an instant Instagram story:

• Planted, graffitied motor cars at Cadillac Ranch, Route 66, Amarillo, Texas
• Area 51 warning signs on the Extraterrestrial Hwy, State Route 375, Nevada
• Candy-coloured messages of hope at Salvation Mountain, Niland, California

• Teepee-shaped accommodation at the Wigwam Motel, Holbrook, Arizona
• The pint-sized Eiffel Tower, Lady Liberty and Sphinx on The Strip, Las Vegas, Nevada

NATIVE AMERICAN CULTURE

The USA's indigenous culture has not been heavily promoted but there are numerous sites where you can connect with Native American culture for free, or almost free, though many choose to use the services of local guides to direct money to indigenous communities.

Canada has done a rather better job of channelling tourist income to First Nations communities. Investigate the community tourism initiatives at Six Nations Grand River (sixnationstourism.ca; Brantford, Ontario) and Great Spirit River Trail (circletrail.com; M'Chigeeng, Ontario).

Below: Thunderbird Park in Victoria, Canada, features totem poles restored by Kwakwaka'wakw master carver, Mungo Martin.

Central America has its own rich pre-Columbian traditions, best experienced in Oaxaca in Mexico and along the Ruta Maya, which connects several countries and along which Maya culture still thrives.

TOP NATIVE AMERICAN SITES

To gain a glimpse of the Americas that existed before the continent was named, visit the following sites:

• Petroglyph National Monument (nps.gov/petr, Albuquerque, New Mexico; free) – one of the US' largest collections of pre-Colombian petroglyphs.

• Canyons of the Ancients (blm.gov; Cortez, Colorado; free) – ancient homes of Pueblo people in a wild desert landscape.

• Effigy Mounds National (nps.gov/efmo; Harpers Ferry, Iowa; free) – zoomorphic earthworks created by people from the Iowa, Sioux and Sac and Fox nations.

• Writing-on-Stone Provincial Park (albertaparks.ca/parks/south/writing-on-stone-pp; Lethbridge, Alberta, Canada; free) – First Nations rock carvings in a protected prairie.

Above: visiting small towns, for example along Route 66, will reveal some bargain sights and attractions.

HIGHLIGHTS

SMITHSONIAN MUSEUMS, WASHINGTON, DC

With museums covering every aspect of US history and culture, from aviation and space travel to dinosaurs and geology, and from the Constitution to African-American heritage, the Smithsonian is the nation's greatest cultural resource, and entry to every institution is free. *p174*

PERFORMANCES AT CENTRAL PARK, NEW YORK

New York's most famous park is the setting for a great programme of free summer theatre, music and dance, from legendary Shakespeare plays at the open-air Delacorte Theater to world music and contemporary dance care of Summerstage. *p176*

GETTY CENTRE, LOS ANGELES

The collection founded by petro-industrialist Jean Paul Getty is housed in a work of art itself – a billion-dollar Modernist perch poised high above LA – but against all odds, entry is free. Inside, you can feast your eyes on works from Vincent van Gogh's *Irises* to *Arii Matamoe* by Paul Gauguin. *p178*

PALACIO NACIONAL, MEXICO CITY

It's not the palace that's the attraction, though the grand colonial-era structure is certainly impressive. What matters is what's inside: Diego Rivera's legendary murals of Mexican civilisation, tracing centuries of history from Aztec legends to revolution and beyond. *p179*

NORTH AMERICA

WASHINGTON DC, USA

*Washington DC leads the world in free entertainment. The cost of museums? Gratis.
Concerts? Nada. Visits to the White House and Congress? Zilch. Even drinking venues
throw in promo beverages. You can have days of fun without spending a dime.*

NORTH AMERICA

Smithsonian Museums

DC surely boasts the world's highest number of free-entry museums, thanks to the Smithsonian Institution – a world-leading collection of 19 museums and galleries, plus a zoo and research centres. Visitors can learn about virtually everything, from dinosaurs to space shuttles, gold nuggets to the Star-Spangled Banner and, most recently, African-American culture care of the National Museum of African American History & Culture. *si.edu; free.*

Bureau of Engraving and Printing

Just in case (with all these freebies) you've forgotten what money actually looks like, head here to see US dollars being printed. Free 40-minute tours visit the production floor where you can watch millions roll off the presses. *moneyfactory.gov/ washingtondctours.html; 14th & C Sts, SW; closed Sat & Sun; free.*

Library of Congress

Not only does it hold one of the world's largest collections of books, but the stunning Thomas Jefferson Building here is replete with mosaics, paintings and fascinating history. Regular free tours give a rundown of the many chapters of this palace of pages. *loc.gov; 101 Independence Ave SE; closed Sun; free.*

National Archives

This is the place where locals' hearts go aflutter. The reason? It houses the country's Constitution, the Bill of Rights and the Declaration of Independence. Also dip into the public vaults, a fascinating, interactive collection of original records from Abraham Lincoln's telegrams to recordings from the Oval Office. *archives.gov/dc; 700 Pennsylvania Ave NW; closed Sat & Sun; free.*

National Theatre

Children can get their free fun on Saturday mornings at National Theatre events that inspire creativity and imagination through play, puppets, interactive performances, dance and

Below: inside the John F Kennedy Center for the Performing Arts. Right: the interior of the Library of Congress.

music. Tickets are distributed on a first-come-first-served basis 30 minutes before the curtain rises. *thenationaldc. com/#community-programs; 1321 Pennsylvania Ave NW; Sat; free.*

✋ Pentagon tour

Make reservations from 14 to 90 days in advance for a glimpse inside the US Department of Defense headquarters. Tours cram a lot into the hour: an informative run-down on the military, visits to memorials and plenty of unusual facts and figures, while you explore accessible parts of this fortress-like building. *pentagontours.osd.mil/Tours; The Pentagon; closed Sat & Sun; free.*

✋ Jazz at the National Gallery's Sculpture Garden

Swing by the National Gallery of Art Sculpture Garden on summertime Friday evenings to catch free performances of jazz, salsa, Afrofunk and more. DC is a bastion of jazz – greats such as Duke Ellington and Shirley Horn cut their teeth here – and these al-fresco soirées make it available to the masses. *Constitution Ave NW, btwn 3rd & 9th Sts; Fri May-Sep; free.*

✋ John F Kennedy Center for the Performing Arts' Millennium Stage

Every evening, you can catch a live show at the Millennium Stage of DC's performing arts memorial. Acts vary from the sublime (Washington National Opera youngsters) to the ridiculous (skateboarders improvising to music). The best-in-show award goes to the setting on the banks of the Potomac River. *kennedy-center.org/programs/ millennium; 2700 F St NW; free.*

✋ DC cultural tours

You can pound the pavements on your own using the free maps, apps and audio

of Cultural Tourism DC (culturaltourismdc. org). For a more local flavour, head off on a day- or night-tour with a guide from DC by Foot (freetoursbyfoot.com/washington-dc-tours); themes range from secrets and scandals to a straightforward National Mall stroll. *Venues & times vary; tip appreciated.*

✋ Meander the National Mall's monuments

The Mall – a 5km-long rectangle of patchy grass – is 'America's front yard'. Anchored by the US Capitol at one end and the Lincoln Memorial at the other, the strip is dotted with other memorials, including monuments for Vietnam veterans and Martin Luther King, Jr. Be sure to ascend the Washington Monument (also free) for a bird's-eye view. *900 Ohio Dr SW; free.*

✋ The National Gallery of Art

At Washington's most important art museum, modern sculptures fill the Sculpture Garden; the neoclassical West Building showcases European art through to the early 1900s; and the IM Pei-designed East Building displays contemporary art. Look for works by Manet, Monet and Van Gogh, Jackson Pollock and Picasso. *nga. gov; 6th St and Constitution Ave NW; free.*

THE SEATS OF POWER

Incredibly, you can experience the three branches of the US government – Executive (the White House), Legislative (US Capitol) and Judicial (the Supreme Court) – in one day. The White House (whitehouse.gov; 1600 Pennsylvania Ave NW) requires booking through your embassy. You can roll up to the US Capitol (visitthecapitol. gov; East Capitol St NE & First St SE) for a same-day ticket, or book online beforehand. The Supreme Court (supremecourt. gov; 1 First St NE) offers glimpses of court in session or you can sit for an entire hearing.

NORTH AMERICA

NEW YORK CITY, USA

NYC is known for its world-class museums and theatres, diverse neighbourhoods and peerless dining scene. It's also pricey – but though this city is no cheap date, there are still ways to save money, from catching free outdoor concerts to gratis days at art galleries.

NORTH AMERICA

American Folk Art Museum
Outsider artists steal the show at this small museum on Manhattan's Upper West Side. You'll find everything from wood carvings to hand-tinted photographs. There's free music on Friday evenings. *folkartmuseum. org; 2 Lincoln Sq; closed Mon & Tue; free.*

Chelsea Galleries
In arty West Chelsea, free-to-visit galleries display works by famous names, and on Thursday nights you can often enjoy free wine at openings. Start your tour at David Zwirner, Gagosian and Barbara Gladstone. *chelseagallerymap.com; West Chelsea; free.*

National Museum of the American Indian
Despite its setting in a beaux-arts Lower Manhattan building, this Smithsonian affiliate often gets overlooked. But don't miss out: the collections include decorative arts, textiles and ceremonial objects that document America's diverse native cultures. *nmai.si.edu; 1 Bowling Green; free.*

Sunny's Saturdays
Near the waterfront in Brooklyn's Red Hook district, this old-school bar serves up beers and bluegrass. There's no charge, though it's polite to drop some dollars in the hat for the musicians. *sunnysredhook.com; 253 Conover St; Sat; free.*

Brooklyn Bridge Park
It's hard not to fall for Brooklyn after a wander through this green space hugging the East River. Admire views of Lower Manhattan, play volleyball off Pier 6, and ride Jane's Carousel near the foot of the Brooklyn Bridge. *brooklynbridgepark.org; free.*

Free shows at Central Park
Spend a summer evening catching a free show in Central Park. Summerstage (cityparksfoundation.org/summerstage; Jun-Aug) presents an incredible line-up of dance and music, while free Shakespeare in the Park (publictheater.org/programs/ shakespeare-in-the-park) shows take place at the Delacorte Theater (May-Aug). *Central Park, entrance near Fifth Ave & 72nd St; free.*

Below: Central Park in Manhattan often hosts free concerts in the summer at several stages or on the Great Lawn.

© Guillaume Gaudet | Lonely Planet

MONTRÉAL & QUÉBEC, CANADA

Québec is the Canada that might have been if the French rather than British had gained the upper hand 250 years ago. In Montréal, Québec City and around, you can tour museums and state monuments for free in this most elegant of Canadian states.

🖐 Musée Redpath, Montréal

A Victorian spirit of discovery pervades this old natural history museum stuffed with, well, stuffed animals, alongside dinosaur skeletons, seashells, Egyptian mummies, shrunken heads and artefacts from ancient Mediterranean, African and East Asian civilisations. *mcgill.ca/ redpath; 859 Rue Sherbrooke Ouest; by donation.*

🖐 Parc du Mont-Royal, Montréal

Montréalers are proud of their eponymous 'royal mountain,' landscaped by New York Central Park designer Frederick Law Olmsted. In fine weather, enjoy panoramic views from the Belvédère Kondiaronk fronting Chalet du Mont-Royal, a grand stone villa that hosts big-band concerts in summer. *lemontroyal.qc.ca; 1260 Chemin Remembrance; free.*

🖐 Musée des Beaux-Arts de Montréal

The Museum of Fine Arts has amassed centuries' worth of international artworks, but the collection really shines when it comes to Canadian artists such as Prudence Heward, Paul Kane and the Group of Seven. Entry is free on the first Sunday of the month. *mbam.qc.ca; 1380 Rue Sherbrooke Ouest; closed Mon; entry C$16-24, free for under 20s and first Sun of month.*

🖐 Basilique Notre-Dame, Montréal

Built in 1829, the landmark Notre Dame Basilica is a visually pleasing if slightly gaudy symphony of carved wood, paintings, gilded sculptures and stained-glass windows. It was the site of the state funeral of Pierre Trudeau, and Celine Dion's wedding. *basiliquenotredame.ca; 110 Rue Notre-Dame Ouest, Montréal; entry C$10, free during services.*

🖐 Hôtel du Parlement, Québec City

Home to Québec's Provincial Legislature, the gargantuan, statue-crammed Parliament building was completed in 1886. Free 30-minute tours get you into the National Assembly Chamber, Legislative Council Chamber and President's Gallery. *assnat.qc.ca/en/ visiteurs; 1045 Rue des Parlementaires; Montcalm & Colline Parlementaire; free.*

🖐 Basilique Ste-Anne-de-Beaupré, Ste-Anne de Beaupré

The Québec village of Ste-Anne de Beaupré is renowned for its Goliath-sized basilica, a lavish 1920s construction over the site of a succession of earlier churches dating back as far as the 1600s. Inside, don't miss the stunning stained-glass windows and glittering ceiling mosaics depicting the life of St Anne. *sanctuairesainteanne.org; 10018 Ave Royale; free.*

LOS ANGELES, USA

Venture beyond LA's starry-eyed Hollywood glamour to world-class cultural institutions, beaches, hikes and a mini-UN of ethnicities. Even if this is one of America's most expensive megacities, being a culture vulture needn't cost a cent.

🖐 The Broad Museum

The architecture is as cool as the art at this gallery displaying the collections of billionaire philanthropists Eli and Edythe Broad. A who's who of the postwar art world is stunningly displayed behind perforated exterior walls collectively nicknamed 'the veil'. *thebroad.org; 221 S Grand Ave; closed Mon; free.*

🖐 Getty Centre

In its billion-dollar, in-the-clouds perch, this marvel of a museum has quadruple delights: a stellar art collection (from Renaissance greats through to David Hockney); cutting-edge architecture; seasonally changing gardens; and kingly views across the LA Basin. *getty.edu; 1200 Getty Center Dr, Westside; closed Mon; admission free, but parking US$15.*

🖐 LA Philharmonic buildings

Architect Frank Gehry pulled out all the stops for the gravity defying Walt Disney Concert Hall. If you can't fork out for an LA Phil concert, explore on a free self-guided audio tour. In summer, the Phil shifts to the Hollywood Bowl (hollywoodbowl.com; 2301 Highland Ave), the cheapest seats cost just US$1. *laphil.org; 111 S Grand Ave; see website for tickets.*

🖐 Saturday night at the movies

Savvy Angelenos love offbeat movie screenings. The Last Remaining Seats Film Series (laconservancy.org) screens classics in theatres from cinema's golden age. Cinespia (cinespia.org) has various locations, including Hollywood Forever Cemetery (6000 Santa Monica Blvd), eternal resting place of film legends. *Times & prices vary.*

🖐 Step out on Hollywood Boulevard

Marilyn Monroe, Charlie Chaplin, Aretha Franklin and Big Bird are among the 2400 stars sought out, worshipped and stepped on along the Hollywood Walk of Fame (walkoffame.com; Hollywood Blvd; free). If you take a picture with a costumed character, tip a coupla bucks or face wrath.

🖐 Artwalk Gallery Tours

A mad swirl of art lovers invades the city centre for monthly Downtown art walks: self-guided, liberally lubricated excursions that link more than 40 galleries and museums, most between 3rd and 9th and Broadway and Main. *downtownartwalk.org; 2nd Thu of month; free.*

MEXICO CITY, MEXICO

Cultured, culinary and constantly creative – particularly during the annual Day of the Dead celebrations – Mexico City is life in living colour. Much is free to enjoy: you won't need Aztec gold to make the most of the Mexican capital.

✋ Secretaría de Educación Pública

The two front courtyards at this government compound are lined with 120 fresco panels painted by Diego Rivera in the 1920s – an evocative tableau of 'the very life of the people,' in the artist's words. Get up close to see every detailed brushstroke. *República de Brasil 31; closed Sat & Sun; free.*

✋ Palacio Nacional

Inside this colonial palace, Diego Rivera imagined Mexican civilisation from the arrival of Quetzalcóatl (the Aztec plumed serpent god) to the post-revolutionary period. It houses the bell rung in the town of Dolores Hidalgo to start the War of Independence. *historia.palacionacional.info; Plaza de la Constitución; closed Mon; free.*

✋ Palacio de Bellas Artes

Murals by Mexican artists dominate the top floors of this white-marble palace – a concert hall and arts centre commissioned by President Porfirio Díaz. It's free on Sunday. *museopalaciodebellasartes.gob.mx; Av Juárez; closed Mon; entry M$70, free Sun.*

✋ People-watch at Zócalo

The heart of Mexico City, the Plaza de la Constitución – aka Zócalo – is the best place to see Mexico City life unfold. During Día de Muertos in November, the square fills with altars. *Plaza de la Constitución; free.*

✋ Museo Jumex

Free on Sundays, stylish Museo Jumex was built to house one of Latin America's leading contemporary art collections. The Jumex has featured works from renowned Mexican and international artists including Gabriel Orozco, Francis Alÿs, Andy Warhol and Jeff Koons. *fundacionjumex.org; Blvd Miguel de Cervantes Saavedra 303; closed Mon; entry M$50, free Sun.*

✋ Free movies at Cineteca

Mexican and foreign indie movies are shown at the architectural gem of Cineteca in Coyoacán. From October to March catch free open-air screenings at dusk in the rear garden. *cinetecanacional.net; Ave México-Coyoacán 389, Colonia Xoco; free.*

Below: Day of the Dead brings colour and costumes to Mexico's streets, squares and cemeteries.

NORTH AMERICA

TOP

In search of budget thrills and spills in the land of the free and home of the brave? Look no further.

SAN FRANCISCO PARKOUR, CALIFORNIA

It's hard to keep flowers in your hair during a monkey vault, but whether you're a trained *traceur* or a free-running virgin, the SF parkour scene is both developed and welcoming. Complimentary intro sessions offered. *sfparkour.com; free.*

CANYONEERING, UTAH

A non-technical introduction to an addictive art, the spectacular Peekaboo–Spooky Gulch Loop and Escalante-Grand Staircase National Monument in Utah can be spliced into easy 5km return scrambles, with plenty of slots and arches to explore. *utah.com/hiking; free.*

TRAIL-RUNNING, NEW YORK

In autumn, go trail-running around Lake Placid, NY, and explore the tracks that wend through vast, fantastic forests in the Adirondack Mountains to experience an explosion of leaf-turning colour. *lakeplacid.com; free.*

BOULDERING, CALIFORNIA

Another essentially equipment-free pursuit, bouldering is all about solving small climbing problems on, yep, boulders. Bishop, in California's lower Sierra Nevada Mountain Range, is one of the world's best bouldering destinations. *bishopvisitor.com; free.*

LIFE AQUATIC, FLORIDA

Armed with fins and a snorkel set, explore Florida Keys National Marine Sanctuary, which protects the planet's third-largest reef and contains the submerged statue *Christ of the Abyss*. *floridakeys.noaa.gov; free.*

© Stephen Frink | Getty Images

GLACIER-SPOTTING, ALASKA

In summer, the roadside Child's Glacier near Cordova, Alaska, sees a collapse of ice every 15 minutes. In winter, head to the Mendenhall Glacier, a popular walking stop with free winter entry, accessible by shuttle bus from Juneau ($8). *traveljuneau.com; Oct-Apr; free.*

BODY-SURFING, HAWAI'I

No board? No money? No worries! Point Panic beach in Hawai'i, in the midst of a surfing mecca, has a wave so tailor-made for body-surfing that boarders stay away, leaving it to the penny-pinching purists. *Hawai'i, HI; free.*

DEEPWATER SOLOING, UTAH

DWS is rock climbing without the encumbrance (and expense) of safety gear – it's just you, the rock and the water below. Lake Powell's extensive shoreline in Glen Canyon National Recreation Area is the perfect spot to give it a try. *nps.gov/glca; free.*

STORM-CHASING, OREGON

The self-proclaimed 'storm-watching Capital of the World', Bandon-by-the-Sea's beaches become wind- and wave-whipped once winter riles the Pacific into a rage. Watch the action from behind the sea stacks and spires on Bullards Beach. *Free.*

NORDIC SKIING, MICHIGAN

Forget expensive lift passes, extortionate accommodation rates and crowded runs, and try your arms (and legs) at Nordic skiing somewhere like Higgins Lake, Michigan, where you can ski a groomed 18-km network of trails for just $9 per day. *crosscountryski.com.*

ALL-AMERICAN ADVENTURES

© Lindsay Lauckner Gundlock | Lonely Planet

FOOD & DRINK

If we were just talking volume, North America would be one of the world's greatest destinations for money-conscious gourmets. From street-side taco feasts in Mexico to monster burgers in the US, portion sizes can be huge, so mix things up with healthy foods from farmers' markets, and save the expensive sit-down meals for local specialties you can't find elsewhere.

SELF-CATERING & FOOD MARKETS

Self-catering in America and Canada

Above: Mexican food, like these shrimp enchiladas from the Pacific coast, is delicious, nutritious and often great value.

frees you from the tyranny of fried food and having to dig deep for tips. From Anchorage to Zapopan, local food markets are on hand to provide you with the finest produce and delicacies at fair prices (with bonus tropical fruit in Central America and the Caribbean).

• **Cook out** – with a camping stove, every campsite becomes a cordon bleu kitchen; and many sites have free barbecues for campers, for cost-friendly meat feasts.

• **Hire an RV** – you'll pay more than tents at campsites, but not much more, and the in-van kitchen is great for more ambitious dinners.

• **Hail the kitchenette** – plenty of motels offer rooms with a simple kitchenette; being able to keep things cold opens up all sorts of creative cooking options.

FAST FOOD

Things have come on a long way since Richard and Maurice McDonald opened their burger stand in San Bernardino, California in 1940. Fast food today encompasses everything from bicycle-borne tamales in Mexico City to kimchi-topped burgers served from San Francisco food trucks. You won't have to hunt to find street food feasts at a bargain price.

• **Go local** – the Americas' best food-truck hangouts are often not found downtown but in the suburbs, where vacant lots, street markets, haboursides and public parks are transformed into gourmet hubs.

• **Make for the markets** – farmers' markets and public produce markets are fabulous places to graze, with plenty of cooked-food stalls supplementing the piled local produce.

• **Small treats** – in Mexico and across Central America, look for *antojitos* (small cravings) – cheap street snacks that can easily be assembled into a tasty and filling meal.

SIT-DOWN DINING

As soon as there's table service in the US and Canada, prices creep up, not least because of the obligatory tip (15-20% is the norm). For bargains, head to smalltown diners, or migrant neighbourhoods and studenty quarters, where culinary creativity comes at a (relatively) moderate price.

• **Believe in the buffet** – all-you-can-eat buffets are found all over North America, offering volume and budget prices, sometimes as a trade-off for quality.

• **Big breakfasts** – start the day with a big breakfast and you can cruise through lunch with just a sandwich.

• **Eat the dish of the day** – lunch specials are easy to find, undercutting most dishes on the à la carte menu.

SMART DRINKING

The US' strict licensing laws put drinking at any price out of reach of the under 21s (under 18s in most of Canada), and widespread 'open container' laws put the kibosh on public drinking, even in the privacy of your own vehicle. Drinking out is more relaxed in Central America and the Caribbean...and cheaper if you stick to beer and excellent local rums.

The USA's wine regions offer plenty of tasting opportunities, but there's often a meaty charge to sip. Trade Napa and Sonoma for cheaper wine country in the Finger Lakes (fingerlakes.org/things-to-do/wineries), the hill country of Texas, Canada's Niagara (visitniagaracanada. com/taste/wineries) and Okanagan Valley (winebc.com/discover-bc-wine-country/okanagan-valley), or Mexico's Valle de Guadalupe.

HIGHLIGHTS

FERRY PLAZA FARMERS MARKET, SAN FRANCISCO
Join SF foodies on a gastronomic world tour at the Ferry Building market, where fusion foods and rotisserie sandwiches supplement the fresh, nutritious Californian produce on all sides. Free samples abound and you can haul off your finds for a picnic by the bay. *p186*

MISTER GREGORY'S, NEW ORLEANS
Get your Creole on at this legendary no-frills breakfast and lunch spot, where Francophiles gather for deli baguettes, melted cheese and béchamel sandwiches, and old-fashioned New Orleans-style shrimp boils. *p189*

NORTH AMERICA

It's CHURRO Time!

TRADITIONAL
- Cinnamon Sugar Churros $3.00
    - Chocolate Dipping Sauce + .50¢
    - Cajeta Dipping Sauce + .50¢

$4.50 **TEXAS COMFORT** ★
- Cinnamon Sugar
- Apple Pie Filling
- Caramel Sauce
- Whipped Cream

$4.50 **date Night** »»→
- Brown Sugar
- Salted Caramel
- Vanilla Ice Cream

$4.50 **CAMP FIRE** ⬆
- Graham Sugar
- Toasted Marshmallow
- Mexican Chocolate Sauce
- Whipped Cream

$4.50 **Wimberley** ☁
- Sage Sugar
- Blackberry Jam
- Whipped Cream
- Fresh Flowers

$4.50 **Rico Suave** ♡
Cocoa Sugar / Strawberry Jam / Nutella / Caramel / Whipped Cream

★ **Add A Scoop of Blue Bell Vainilla Ice Cream $1.00**

BEST STREETFOOD

The US caught the street-food bug early and food trucks cluster around city centres, but the rest of the continent also offers delicious variants of on-the-go-snacking.

THE ALL-AMERICAN HOT DOG, DENVER, USA

Served from food carts around Denver and a brick-and-mortar premises on Latimer St, Biker Jim's Dogs was the brainchild of a former car repo guy, who turned his hand to making Denver's finest hot dogs. Go classic, or off-piste with rattlesnake and pheasant sausage. *bikerjimsdogs.com; 2148 Larimer Street, Denver; hot dogs US$7.50.*

STREET TACOS, SAN FRANCISCO, USA

San Francisco does tacos right, the real Mexican way. At Mariscos Jalisco's food truck (facebook. com/mariscosjalisco; 3040 E Olympic Blvd), stunning shrimp tacos are the main attraction, while Leo's Tacos Truck (leostacostruck.com; seven downtown locations) sticks to the classics: tacos *al pastor*, with spit-grilled pork. *Tacos from $2.50.*

MAC AND CHEESE, PHILADELPHIA, USA

Mac Mart has moved on from the days when its legendary big pink truck used to park up at Drexel University, but you can still get Philly's favourite creamy mac and cheese to go at this bricks-and-mortar eatery on S 18th, including stuffed into a sandwich as the 'Return of the Mac'. *macmartcart. com; 104 S 18th St, Philadelphia; from US$8.*

LOBSTER ROLLS, NYC, USA

Founded by a former investment banker, Luke's Lobster has grown into an empire that spans the US and Asia. Their low-key FiDi food shack is still the best place to come for finger-licking lobster, crab and shrimp rolls with fries or slaw – a splurge but a treat. *lukeslobster.com; 26 S William St, New York; rolls and fries from US$17.*

POKE BOWL, HONOLULU, USA

Hawai'ian *poke* – raw tuna, salt, soy, sesame oil, seaweed and chili – is an island institution, revealing Hawaii's fusion leanings. You'll find it across the islands, but locals have a special love for the bowls served to go at Honolulu's Ono Seafood. Buy by the pound or with rice as a *poke bowl. 747 Kapahulu Ave, Honolulu, O'ahu; poke from US$7.*

FUKUBURGER, LAS VEGAS

When it comes to burgers, the only rule is there are no rules. Las Vegas's famous Fukuburger updates the American version with Japanese wasabi mayo and teriyaki sauce. Find their food truck at events all over Vegas, or visit their restaurants at 7365 S Buffalo Lane and 3429 S Jones. *fukuburger.com; burgers US$8.*

JERK CHICKEN AT ANDY'S, KINGSTON, JAMAICA

If you're after authentic jerk chicken and pork, this modest corner stop is a local legend. It gets particularly busy in the evenings, when locals line up for their meats accompanied by fried breadfruit, festival (fried dumplings), sweet potato or plantain. *49 Mannings Hill Rd; closed Sun; jerk chicken from J$600.*

POUTINE AT LA BANQUISE, MONTRÉAL, CANADA

This street treat of chips smothered in gravy, cheese and other toppings pops up everywhere, but the portions served at 24-hour diner La Banquise are some of the best in the country; try the vegan and pulled-pork varieties. *labanquise.com; 994 Rue Rachel Est; 24hr; poutine from US$8.*

STREET TAMALES, MEXICO CITY, MEXICO

Cycle-riding vendors wake up Mexico City every morning with the word 'Tamales!' blaring from speakers as they tout their wares. These stuffed snacks, wrapped in corn dough and steamed in leaves, are the fuel of Mexico. *Tamales around M$7.*

Left: in cities such as Austin, Texas, food trucks offer inventive and affordable food with a sociable vibe on the side.

NORTH AMERICA

NORCAL, USA

From flower-power to Yosemite, Northern California – NorCal to its friends – has long captured America's imagination. Foodie entrepots such as San Francisco and Oakland are a pocket-friendly feast, and even the winelands don't have to break the bank.

Ferry Plaza Farmers Market, San Francisco

The pride and joy of SF foodies, the Ferry Building market showcases more than 100 prime purveyors of California-grown organic produce, pasture-raised meats and gourmet prepared foods at accessible prices. Graze, then haul away some goodies for a bayside picnic. *cuesa. org; cnr Market St & the Embarcadero; Sat, Tue & Thu; street food from US$3.*

Off the Grid, San Francisco

Spring through fall, food trucks and pop-up cubes circle their wagons at SF's largest mobile-gourmet hootenannies on Friday night at Fort Mason Center, and for brunch on Sunday for Picnic at the Presidio on the Main Post Lawn. Arrive early for the best selection and to minimise waits. *offthegridsf.com; 2 Marina Blvd, Fort Mason Center; food from US$6.*

Hook Fish Co, San Francisco

There's a reason you packed that warm coat: to visit this delightful fish joint in windy Outer Sunset. Order over a small wooden counter from a short but enticing menu of locally sourced seafood, including a *poke* burrito, trout salad and an array of fish-of-the-day tacos. *hookfishco.com; 4542 Irving St, San Francisco; mains from US$13.*

Golden Boy, San Francisco

'If you don't see it don't ask 4 it' reads the menu – Golden Boy has kept punks in line since 1978, serving Genovese focaccia-crust pizza that's chewy, crunchy and hot from the oven. You'll have whatever the Sodini family are making and like it – especially pesto and clam-and-garlic. *goldenboypizza.com; 542 Green St; pizza slices from US$3.25.*

RT Rotisserie, San Francisco

An all-star menu makes ordering mains easy – you'll find bliss with entire chickens hot off the spit, succulent lamb and pickled onions, or surprisingly decadent roast cauliflower with earthy beet-tahini sauce – but do you choose porcini-powdered fries or salad with that? *rtrotisserie.com; 101 Oak St; dishes from US$9.*

Good Mong Kok, San Francisco

Ask Chinatown locals about their go-to dim sum and the answer is likely to be either grandma's or Good Mong Kok. Join the line outside this counter bakery for dumplings – classic pork *siu mai*, shrimp *har gow* and BBQ pork buns – which are whisked from vast steamers into takeout containers to enjoy in Portsmouth Sq. *1039 Stockton St; dumplings US$2-5*

Below: buy organic produce at the Grand Lake Farmers Market in Oakland; there's also a regular Saturday market in nearby Berkeley. Right: food trucks, such as the Chairman, proliferate in San Francisco.

🌐 Oakland–Grand Lake Farmers Market, Oakland

A rival to San Francisco's Ferry Plaza Farmers Market, this bountiful weekly market hauls in bushels of fresh produce, ranched meats, artisanal cheese and baked goods from as far away as Marin County and the Central Valley. Food stands throng – don't skip the dim-sum tent. *splashpad.org/farmers-market; Lake Park Ave, at Grand Ave, Oakland; Sat; snacks from US$3.*

🌐 Taqueria El Paisa@.com, Oakland

Look for a low-slung little restaurant with a green awning, popular with Oakland's Mexican population, and strap in: it's not just excellent, it's dirt cheap. The tacos are good beyond all superlatives: simple, fresh and garnished with coriander, onions and *nopales* (cactus). *facebook. com/pg/elpaisa77; 4610 International Blvd, Oakland; tacos from US$3.*

🌐 Oxbow Farmers' Market, Napa

Why commit to just one dining establishment when you could graze a dozen of Napa's finest for a fraction of the cost? Assemble the meal of your California dreams with all-star dishes – perhaps Hog Island Oyster Co oysters mignonette with a Fieldwork Brewing Company farmhouse ale. *oxbowpublicmarket.com; 610 & 644 1st St, Napa; snacks from US$3.*

🌐 Locals Tasting Room, Sonoma

Getting a free wine sample in the Napa Valley requires god-like powers of persuasion. Over in Sonoma, things are more chilled, particularly at this independent, collective tasting room – the first in California. Try before you buy without the compulsion to make a purchase. *localstastingroom.com; 21023 Geyserville Ave, Geyserville, Sonoma County; hours vary.*

🌐 Spud Point Crab Company, Bodega Bay

In the classic tradition of dockside crab shacks, Spud Point serves salty-sweet crab sandwiches and real clam chowder (that consistently wins local culinary prizes). Get it to go, or eat at picnic tables overlooking the marina. *spudpointcrabco. com; 1910 Westshore Rd, Bodega Bay; mains from US$8.*

🌐 Davis Farmers Market, Davis

With more than 150 vendors and an awe-inspiring selection of local produce, this is justly considered among the best farmers' markets in the nation, and it's great for low cost munching. On Wednesday evenings in summer, gather provisions for a dreamy dinner picnic in the park. *davisfarmersmarket.org; cnr 4th & C Sts, Davis; Wed & Sat; snacks from US$3.*

ASIA IN AMERICA

The West Coast is well known for its Chinatowns, founded by Chinese miners and railroad workers in the 19th century, and famous today for cheap and highly authentic Asian food. The name 'Chinatown' was first coined by US newspapers in 1853 for San Francisco's Chinese quarter, centred on Grant Avenue and Stockton Street, which still throngs with cheap roast duck houses, noodle shops and weekend dim sum canteens; bring an appetite, not a fat wallet.

NORTH AMERICA

OAXACA, MEXICO

Authentic Mexican food is varied and very tasty, and there's nowhere better to undertake a culinary expedition than southern Mexico's Oaxaca region, where delicious food comes with a side of surf breaks, indigenous culture, colonial cities and forested mountains.

✋ Zandunga, Oaxaca City

Zandunga brings the taste of Tehuantepec to the capital: succulent meats and seafood combined with tropical fruits, and dishes cooked in banana leaves. It's worth the small investment for the *botana zandunga* (a sampler of Oaxacan dishes which easily serves two). *zandungasabor.com; C/García Vigil 512E, Oaxaca City; mains from M$90.*

✋ Boulenc Pan Artesano, Oaxaca City

At Oaxaca's best bakery cafe, the most popular dish is avocado toast (but hey, avos come from here). Equally excellent coffee is served in terracotta mugs. *facebook.com/Boulenc.oax; C/Porfirio Díaz 207, Oaxaca City; closed Sun; mains from M$54.*

✋ Zapoteca Cocina del Istmo, La Crucecita

For smoking hot regional food on the Pacific Coast, head to this intimate spot, where fair prices get you feasts such as *garnachas istmeñas* (fried tortillas with shredded beef, cheese and pickled cabbage). *C/Flamboyán 211, La Crucecita; mains from M$70.*

✋ La Tosta, Santa Cruz Huatulco

If you can find it (follow the hungry locals by Hotel Binniguenda), this is the place to track down artfully prepared tostadas piled high with fresh tuna, octopus, shrimp and ceviche. *facebook.com/latostahuatulco;*

Blvd Juárez, Santa Cruz Huatulco; closed Mon; mains from M$45.

✋ El Cafecito, Puerto Escondido

This lively spot serves gringo-Mexican fusion food with surfer appetites in mind. Some of the best breakfasts on the coast include spiced-up eggs and bowls of Oaxacan chocolate. *Ave del Morro, Puerto Escondido; breakfasts from M$59.*

✋ Orale! Café, Zipolite

Breakfast at this garden cafe feels like eating in a jungle clearing. Ask for a fruit plate that might have been picked from a nearby tree (served with yoghurt and granola). *Off Ave Roca Blanca, Zipolite; closed Tue & Wed; breakfast from M$50.*

Below: a low-carbon street snack vendor in Oaxaca.

© Justin Foulkes | Lonely Planet

NORTH AMERICA

TEXAS & LOUISIANA GULF COAST, USA

Austin, Houston and the Gulf Coast of Texas and Louisiana are the places to come to find food that carries a genuinely American stamp – it's usually meaty, sizzling, and served in lavish portions, but not always at lavish prices.

Chuy's, Austin
Millions of students can't be wrong: satisfy your Tex-Mex cravings at the free happy-hour nacho car (weekdays 4-7pm) of this Austin-founded chain. Etiquette dictates you order a beer or margarita. *chuys.com; 1728 Barton Springs Rd, Austin, Texas; free.*

Franklin Barbecue, Austin
This famous BBQ joint only serves lunch, and only until it runs out. Join the line by 10am and bring beer or mimosas to share and make friends. And yes, you do want the fatty brisket. *franklinbbq.com; 900 E 11th St, Austin. Texas; closed Mon; sandwiches from US$7, meat per lb from US$19.*

Amy's Ice Creams, Austin
It's not just the ice cream we love; it's the toppings that get pounded and blended by staff wielding a metal scoop in each hand. *amysicecreams.com; 1012 W 6th St, Austin, Texas; ice cream from US$3.25.*

Lankford Grocery, Houston
This vintage neighbourhood grocery store is now dedicated to serving some of Houston's finest burgers, juicy and loaded with condiments – what more could you want? The interior is shambolic but there's outdoor seating too. *facebook.com/lankfordgrocery; 88 Dennis St, Houston, Texas; closed Sun; burgers from US$7.*

Johnson's Boucanière, Lafayette
This 80-year-old smoker business turns out detour-worthy *boudin* (Cajun sausage), an unstoppable smoked-pork-brisket sandwich topped with sausage, and pulled pork stuffed into grilled-cheese biscuits. *johnsonsboucaniere.com; 1111 St John St; Lafayette, Louisiana; closed Sun & Mon; mains from US$4.25.*

Mister Gregory's, New Orleans
Beloved by French expats, this no-frills breakfast and lunch spot specialises in deli baguettes, croque-style sandwiches (with melted cheese and béchamel on top) and classic New Orleans shrimp boils. *mistergregorys.com; 806 N Rampart St; New Orleans, Louisiana; mains from US$5.*

Below: a filled baguette makes for a quick breakfast or lunch in Texas and Louisiana.

NORTH AMERICA

BOSTON & NEW ENGLAND, USA

Baked beans are way down the list of food treats to come out of New England. This is the home of clam chowder, hasty pudding, succotash (corn with lima beans) and the lobster roll – available at bargain prices in this all-American corner of the country.

Luke's Lobster, Boston

Luke Holden took a Maine seafood shack and plonked it in the middle of Back Bay so that hungry shoppers could get a lobster roll for lunch. The place looks authentic, with a weathered wood interior and nautical decor, but more importantly, the lobster rolls are the real deal – and affordable. *lukeslobster.com; 75 Exeter St, Boston; mains US$9-19.*

Eventide Fenway, Boston

James Beard-award winners Mike Wiley and Andrew Taylor are responsible for this counter-service version of their beloved Maine seafood restaurant. Fast, fresh and fabulous, the menu features just-shucked oysters and sophisticated seafood specials. *eventideoysterco.com; 1321 Boylston St, Boston; mains US$9-16.*

PB Boulangerie Bistro, Cape Cod

A Michelin-starred French baker setting up shop in tiny Wellfleet? Scan the cabinets full of fruit tarts, chocolate-almond croissants and filled baguettes and you'll think you've died and gone to Paris. *pbboulangeriebistro.com; 15 Lecount Hollow Rd, South Wellfleet; closed Tue; pastries from US$3.*

Artcliff Diner, Martha's Vineyard

Culinary Institute of America graduate Gina Stanley adds flair to everything she touches, from almond-encrusted French toast to fresh fish tacos. Easily the best breakfast and lunch in Vineyard Haven. *artcliffdiner.com; 39 Beach Rd, Vineyard Haven; closed Tue & Wed; mains US$9-14*

Helen's Restaurant, Maine

Helen's is the kind of friendly locals' joint where waitresses call you 'hon', but their food makes standard American diner fare look like mud in comparison. Fresh haddock is moist and flaky, and the blueberry pie is the envy of restaurants across the state. *helensrestaurantmachias.com; 111 Main St/US 1, Machias; closed Mon; mains from US$8.*

Shannon's Unshelled, Maine

If you like your lobster rolls simple, with no accoutrements but a dollop of melted butter on the side, Shannon's is for you. The meat is fresh and tender, and it comes on thick, nicely toasted bread. *shannonsunshelled.biz; 11 Granary Way, Boothbay Harbor, Maine; lobster rolls US$7-15.*

NORTH AMERICA

TORONTO, CANADA

A relative newcomer on the world stage, Toronto is becoming increasingly sure of itself, with swanky shops, hip neighbourhoods and gourmet – but still affordable – food, including the offerings of one of North America's great Chinatowns.

♨ House of Gourmet

A busy restaurant specialising in Hong Kong-style *congee* (rice porridge), noodles, barbecue and seafood, more than 800 dishes in fact – it takes two menus to cover all the options. *houseofgourmet.ca; 484 Dundas St W, Chinatown; mains from C$8.*

♨ Parka Food Co

A casual, starkly white restaurant serving delicious vegan comfort food: burgers made from portobello mushrooms and blackened cauliflower; mac 'n' (vegan) cheese with toppings like truffle mushroom, and soups. *parkafoodco.com; 424 Queen St W, Alexandra Park; closed Mon & Tue; mains C$8-13.*

♨ Seven Lives

Formerly a pop-up taqueria, now a hole-in-the-wall place with lines of people waiting to order Baja-style fish tacos: light and flaky mahi-mahi with pico de gallo, cabbage and a creamy sauce. There are other seafood and veggie combos, too. Most diners eat standing or take their meal to nearby Bellevue Square. *sevenlivesto.ca; 72 Kensington Ave, Kensington Market; tacos from C$6.*

♨ Forno Cultura

An Italian bakery tucked into a long room, one side lined with impossible-to-resist goods, the other with a view of bakers doing their thing. Ingredients are imported from the mother country and communal tables encourage you to stay, watch and eat. *fornocultura.com; 609 King St W, Fashion District; baked snacks from C$3.*

♨ Aunties & Uncles

There's usually a queue outside the picket fence of this bustling brunch/lunch joint with a menu of cheap and cheery homemade favourites. Grab one of the mismatched chairs and dig into dishes like grilled brie with pear chutney and walnuts on challah, banana-oatmeal pancakes or grilled Canadian cheddar. *auntiesanduncles. ca; 74 Lippincott St, Harbord Village; closed Sat & Sun; mains from C$8.*

Below: Dundas St in Toronto is at the steamy heart of the city's Chinatown.

© JW_PNW | Shutterstock

NORTH AMERICA

FESTIVALS & EVENTS

Locals insist everything is bigger and better in the US, and when it comes to festivals, they might be right. Enthusiasm is the factor that turns even the smallest county fair into a full-on party here. And across the rest of the continent, from Mexico City to Montego Bay, and Montréal to Miami, there's a rich tradition of free stages, even at the biggest events. Look to festivals organised by city authorities rather than commercial promoters, and you'll hit gratis gold.

Above: take in a live outdoor performance at the Montréal International Jazz Festival.

PRACTICALITIES

North America's biggest festivals see massive relocations of people, meaning transport and accommodation is heavily oversubscribed at most of the big celebrations. Thanks to the influx of tourists from the US and Canada, getting a room for any major festival can be a challenge everywhere in the continent.

• **Beat the rush** – tickets for big music and sporting events sell out within minutes of release; check when tickets go on sale and be ready with a charged phone and credit card.

• **Work it** – most events need short-term help, and throw in free event access for staff. You'll need to able to work locally.

• **Go up-country** – rural festivals are smaller, cheaper, and often much more fun than big-city shindigs, and there's often cheap camping nearby.

• **Head south** – the crush eases slightly once you get south of Mexico, and Panama, Nicaragua and Costa Rica have busy festival calendars.

RELIGIOUS FESTIVALS

If you're after a free spectacle, Halloween is bigger than Christmas. Ghouls, goblins and ghosts fill cities and towns across the US and Canada in late October, and a few days later Mexico offers its own rainbow-coloured spin on All Hallows for the Día de Muertos (Day of the Dead).

Christian feast-days such as Christmas, Easter and Mardi Gras (the last day before Lent) are huge events across North America too, but you'll rack up a bill celebrating Christmas in New York or Mardi Gras in New Orleans. Consider a non-traditional Christmas in the Caribbean, with twinkling lights on the palm trees, and set aside Easter for Guatemala's spectacular Semana Santa (Holy Week).

MUSIC & CULTURE

The US and Canada set the standard for music, cinema, culture and arts festivals. You'll pay a premium for the biggest names, but not always – for every pricey Coachella (coachella.com; Indio, California; Apr) there's a free Chicago Blues Festival (chicago.gov; Chicago; Jun) or Montréal Jazz Festival (Montréal; Jun; montrealjazzfest.com).

• **All arts are not alike** – rock and pop music, theatre and cinema often involve a fee; literature, visual arts, folk music and alternative forms of creativity (poetry, live historical re-enactments) often don't.

• **Celebrate the countryside** – hundreds of county fairs, harvest festivals and local pageants offer cheaper, friendlier fun than the big-city spectaculars.

• **Look south** – Jamaica's annual carnival season sees parties and concerts from December to the main event in Easter.

FOOD

North American food festivals are huge events on the social calendar. As well as pricey, big-city extravaganzas like Taste of Chicago (chicago.gov; Jul) and Dine Out Vancouver Festival (dineoutvancouver. com; Feb), there are hundreds of smaller events celebrating local foodstuffs (clams, ribs, seafood, you name it), where you can feast without breaking out the emergency credit card. In Central America and the Caribbean, fabulous street food accompanies every big event on the festival calendar.

SPORTS

You won't get a ticket to the Superbowl without shedding some serious bucks, but loads of smaller competitions and non-league events charge minimal prices, or nothing at all, in the case of some seasonal sports (p197).

HIGHLIGHTS

DÍA DE MUERTOS, MEXICO
The first days of November see Mexico fill with *altares de muertos* (altars of the dead), coloured sand patterns and sculptures, *comparsas* (satirical fancy-dress groups) and more skull iconography than the rest of the world put together. Mexico's biggest showstopper makes Halloween look like a Sunday school picnic. *p194*

MONTRÉAL INTERNATIONAL JAZZ FESTIVAL, CANADA
You'll be clicking your fingers to complex time signatures aplenty at this huge celebration of world jazz, which pulls in all the biggest names. Best of all, around three-quarters of the shows are free – music to the ears of budget travellers. *p195*

FESTIVALS

Canada Day, Ottawa, Canada

Canadians reach peak patriotic zeal every July when they mark the birth of their nation, with the most boisterous celebrations filling Ottawa, the Canadian capital. Everyone wears red and white, crowds take over the streets, maple motifs abound, and there are fireworks, beer and feasting. *1 July, Ottawa, Canada; free.*

Chicago Blues Festival, USA

Chicago has had the blues for as long as anyone can remember. The city's biggest music jamboree brings out superstars: for three days – free shows fill stages in Grant Park with guitar and vocal virtuosity. *chicago.gov; Jun; Chicago, USA; free.*

Día de Muertos, Mexico

Mexico's Day of the Dead celebrations are best enjoyed in Oaxaca and Mexico City. Homes, cemeteries and public buildings are decorated with fantastic *altares de muertos* (altars of the dead); streets and plazas are adorned with *tapetes de arena* (coloured sand patterns); and *comparsas* (satirical fancy-dress groups) parade the streets. *1-2 Nov; Mexico.*

Festival International de Louisiane, Lafayette, USA

Every April, downtown Layfayette closes to traffic for five days of southern music, spread over eight stages, with proudly Francophone leanings. This is one of the few towns with no law against open alcohol containers, so you can sip a beer while you tap a toe. *festivalinternational. org; Apr; Lafayette, USA; free.*

French Quarter Festival, New Orleans, USA

For the NOLA spirit without the pricetag, come for the city's French Quarter Festival, when 20 free stages host the best of southern music: blues, jazz, Cajun, you name it. Top local singers and musicians put in appearances and there's great festival food to munch while you groove. *frenchquarterfest.org; Apr; New Orleans, USA; free.*

Hardly Strictly Bluegrass, San Francisco, USA

This three-day celebration of homegrown America sounds fills San Francisco's Golden Gate Park with banjos, fingerpicking and fiddles every October, and the foot-stomping comes free. It's one of the few unsponsored musical events in America. *hardlystrictlybluegrass. com; Oct; Golden Gate Park, San Francisco, USA; free.*

Below: marching bands are a highlight of New Orleans' epic Mardi Gras festivities.

Ionia Free Fair, USA

Midwest America's favourite country fair features ten days of parades, circus shows, pig races, livestock competitions, and fairground rides. Some grandstand shows are ticketed, but most events are free; it's a great intro to rural America. *ioniafreefair.com; Ionia, USA; Jul; free.*

New Orleans Mardi Gras, USA

Cheap it ain't thanks to the run on rooms and transport, but most of the spectacles are free at New Orleans' most extravagant party: hallucinogenic floats, masked strangers, marching bands, flying beads and much more. *mardigrasneworleans. com; New Orleans, USA; Feb/Mar; free.*

Montréal International Jazz Festival, Canada

This extravaganza fills Montréal with polyrhythms, syncopation and altered chords every summer, and some three-quarters of the shows are free (though you'll pay to watch the biggest names). *Jun/Jul; Montréal, Canada; most events free.*

Musikfest, Bethlehem, USA

The largest free, non-gated music festival in the US, Musikfest has been rocking Bethlehem, PA, since 1986. It's an eclectic event, spread over 10 days, and past line-ups have included Duran Duran, Sheryl Crow and Kesha. *musikfest.org; July-Aug; Bethlehem, USA; free.*

River to River Festival, New York, USA

Conceived to rebuild the spirit of Manhattan after 9/11, this lively event brings weeks of art shows, concerts, poetry, film screenings and performances to the island of Manhattan. Artists on residencies create special artworks and everything is free to all comers. *lmcc.net/river-to-river-festival; Jul-Sep; NYC, USA; free.*

Semana Santa, Guatemala

Guatemala's biggest celebration paints the streets with colour for the last week of Lent, particularly in Guatemala City and Antigua Guatemala. Watch processions of giant crucifixes and statues, and the creation of 'carpets' in the streets made from coloured sawdust. *Mar/Apr; Guatemala; free.*

Smithsonian Folklife Festival, Washington, USA

America's cultural heritage gets its day in the sun every summer for two weeks around the 4th of July, with traditional arts and crafts and an ethos of inclusivity. *festival.si.edu; Jul; National Mall, Washington, DC, USA; free.*

St Patrick's Day, Chicago, USA

On March 17, the local plumbers' union dyes the Chicago River shamrock-green and a huge parade follows downtown in Grant Park, complete with green hats, marching bands and much consumption of Guinness. *Mar; Chicago, USA; free.*

Yarmouth Clam Festival, USA

Nothing says summertime in Maine like the Yarmouth Clam Festival, when the state's famous clams (beloved by chowder fans everywhere) are shucked, celebrated and consumed in vast numbers. Enjoy parades, fireworks, live-shucking and clam feasts. *clamfestival.com; Jul; Yarmouth, USA; free.*

Carnival, Jamaica

A late arrival to the carnival calendar (first held here in the 1950s), Jamaica's annual spring shindig is a burst of sound and colour. It's also known as Bacchanal, which fits the outrageous costumes, deafening sound-systems and aromatic fragrances. *visitjamaica.com/carnival-in-jamaica; Montego Bay, Ocho Rios and Kingston; some events free.*

ALL HAIL THE COUNTY FAIR!

State and county fairs are as all-American as glazed donuts and apple pie, and they've been filling America's showgrounds since at least the 1800s. Traditionally the focus was on livestock shows, but these days it's all about concerts, fairgrounds and sticky snacks – it's cheaper than a day at Disneyland and usually a lot more fun. As well as the Ionia Free Fair, make time for Texas State Fair (bigtex. com; Dallas; Sep), Minnesota State Fair (mnstatefair. org; Falcon Heights; Sep) and Orange County Fair (ocfair.com; Jul-Aug).

NORTH AMERICA

FREE SPORTS

Sports are a big deal in the US, and prime tickets can cost a fortune – but you can be part of the action and atmosphere for nothing at these athletic alternatives.

US POND HOCKEY CHAMPS, MINNEAPOLIS, MINNESOTA

Simple: if you can stand the cold, you can stand and watch. Minnesota hosts this championship every January, which sees amateur teams do battle on frozen Lake Nokomis. With no seats (or fees), there's nothing between you and the puck-tussling action. *uspondhockey.com; free.*

BOSTON MARATHON, BOSTON, MASSACHUSETTS

The world's oldest annual marathon has been held every Patriots' Day (third Monday in April) since 1897. Standing by the roadside, you're not simply watching 30,000 runners, you're celebrating the history, tradition and – after the bombings of 2013 – a sense of solidarity engendered by all those pounding feet. *baa.org.*

VENICE BEACH STREET-BALL, LOS ANGELES, CALIFORNIA

As well as ogling at the outdoor gym, Venice Beach is the place to see raw, back-to-roots basketball. Street-ball is fluid, free-form basketball, with no ref, fewer rules and lots of showing off. Watch a spontaneous game or come on summer Sundays for the Venice Basketball League.

LITTLE LEAGUE BASEBALL WORLD SERIES, WILLIAMSPORT, PENNSYLVANIA

Tickets for the Baseball World Series can nudge US$1000. Tickets for the Little League version don't cost a dime. This competition has run every August since 1947, and sees players aged 11 to 13 dashing around the Williamsport diamond in front of thousands. *llbws.org.*

IDITAROD, ANCHORAGE, ALASKA

They call this 1600km-long dog-sled dash across Alaska to Nome 'The Last Great Race on Earth®'. Luckily, watching the ceremonial start on the first Saturday in March – from Anchorage's Fourth Ave to Campbell Creek – is less challenging. *iditarod.com.*

EAST COAST SURFING CHAMPIONSHIPS, VIRGINIA BEACH, VIRGINIA

Admire pros and amateurs alike every August at the USA's oldest surfing competition. Since it began in 1963, the event has evolved; it's now a free sports festival, with top skateboarders, skimboarders, stand-up paddleboarders and beach volleyballers on show too. *surfecsc.com.*

SPRING FOOTBALL, COUNTRYWIDE

American Football games aren't cheap. Even College League games cost a pretty penny – unless you catch a practice game in March/April. At these, fans can assess the best new players, grab autographs and get pumped for the upcoming season, all for free. *fbschedules.com.*

NORTH AMERICA

OUTDOORS & ADVENTURE

In North America, the great outdoors is not just great, it's spectacular – huge swathes of the continent are covered by wilderness, from Canada's mighty pine forests and the deserts of the US Southwest to the jungles of Central America, all calling out a siren song to campers, hikers, rock climbers, wildlife spotters and other adventurers.

NATIONAL PARKS & NATURE RESERVES

There's a popular misconception that all of North America's national parks charge

Above: natural wonders like Niagara Falls abound in North America. Right: camping can be a cost-effective way to see the outstanding national parks.

entry fees. In fact, dozens of national parks are a) free to visit and b) just as spectacular as Yosemite, Yellowstone and Banff, but without the crazy crowds.

THE USA

Only 116 of the 421 parks in America charge an entry fee, but that does include some of the most famous. All are administered by the National Park Service (nps.gov), which oversees entry and maintains campgrounds, hiking trails and more.

Fees for the top flight parks hover at around US$30-35 per vehicle and US$15-20 for each person, valid for seven days. Campsite spaces (typically US$15-30 per night) are booked through recreation.gov. The America the Beautiful pass (nps.gov/planyourvisit/passes.htm) offers free entry to all the parks for a year for just US$80.

In addition, there are five designated free days every year, when most parks waive their fees (triggering a bit of a free-for-all, it must be said). Details are posted every year on the National Park Service website.

Don't overlook National Forests and National Grasslands, administered by the US Forest Service (fs.usda.gov); most are free to enter and many offer dispersed camping, but there are modest 'day use' fees (from US$5) for some activities. There are also dozens of free-to-visit National Monuments, managed by the Bureau for Land Management (blm.gov) and other federal agencies.

CANADA, CENTRAL AMERICA & THE CARIBBEAN

Canada's national park service, Parks Canada (pc.gc.ca) broadly follows the US model: each visitor pays a fee for every day inside the park (typically C$10); Discovery Passes are a steal at C$69.19, covering every park for a year. Entry to Central American and Caribbean parks can cost from as little as US$3 per person, but mandatory guide fees for treks can ramp up costs.

GENERAL TIPS

In all national parks in the region, consider the following:
- **National park or national forest?** – many activities can be done in national forests for less, and camping is often free.
- **Seek out state parks** – America's state parks (stateparks.org) and Canada's provincial parks (each state has a website) can rival national parks for natural majesty.
- **Check the fee** – some parks are cheaper than others, some only look cheaper if you ignore camping and transport costs. You only need to visit a few parks for an annual pass to save you a ton of money.
- **Ditch the car** – most national parks charge a fee for cars and RVs; walk or cycle in and you just pay the (smaller) per-person fee.
- **Avoid busy weekends and holidays** – you may not find a camping space inside

WILDLIFE, GUARANTEED

North America's national parks often charge, but costs have to be balanced against the chances of getting close to some amazing creatures. Top spots to score maximum wildlife for your dollar include the USA's Theodore Roosevelt National Park (nps.gov/thro); Teddy Roosevelt came here to roam with bison, and so can you. In Florida's Everglades National Park (nps.gov/ever), self-paddled canoe trips come with definite alligator encounters. And in Canada's Banff National Park (pc.gc.ca/banff), grizzly and black bear are regularly sighted on hikes and dawn or dusk drives.

NORTH AMERICA

Above: explore slot canyons and grander canyons in Arizona.

• **SEE:** Grand Staircase-Escalante National Monument, Utah, USA (entry free)
• **SKIP:** Yellowstone National Park, Idaho/Montana/Wyoming, USA (entry car/person US$35/20)
• **SEE:** Wind River Mountain Range, Wyoming, USA (entry free)
• **SKIP:** Banff National Park, Alberta, Canada (daily entry C$10)
• **SEE:** Kananaskis Valley, Alberta, Canada (entry free)
• **SKIP:** Parque Nacional Volcán Arenal, Costa Rica (entry car/person US$15)
• **SEE:** Parque Nacional Volcán Masaya, Nicaragua (entry from US$4)

ADVENTURE ACTIVITIES

From trekking and climbing to scuba diving and surfing, North Americans make full use of their mountains, deserts, oceans, lakes and plains. This includes hunting, so be aware of hunting seasons and wear bright colours to avoid being mistaken for prey.

With your own gear, one of the world's greatest adventure playgrounds awaits. If you climb, bring your boots. If you surf, grab your board. Gear can be rented locally, but not always inside national park boundaries (if one of those is your destination), and not always cheaply.

• **Walk the beaten path** – in many parks, routes are so well trodden there's little need for a guide – just follow local advice about hiking safely. Note that guides are sometimes mandatory in Central America.
• **Go off-road** – a growing number of parks and reserves are open to mountain bikers, including America's Canyonlands and Big Bend, and Costa Rica's Arenal Volcano.
• **Respect nature** – follow the advice of rangers and locals about hunters, dangerous wildlife and trail hazards; if you get into trouble, the cost of being rescued can be crippling.

the parks at all, and accommodation near the parks vanishes.

NORTH AMERICA'S TOP NATIONAL PARK SWITCHEROOS

Yosemite and Yellowstone draw outdoorsy types like moths to a flame; pick less famous parks with free entry for calmer adventures.

• **SKIP:** Grand Canyon National Park, Arizona, USA (entry car/person US$35/20)
• **SEE:** Canyon de Chelly National Monument, Arizona, USA (entry free)
• **SKIP:** Zion National Park, Utah, USA (entry car/person US$35/20)

© Matt Munro | Lonely Planet

• **The next best thrill** – a scenic flight over Yosemite will cost an arm and leg, but anyone reasonably fit can brave the exposed, cable-assisted hike up the steep slope of Half Dome.

ORGANISED TRIPS

Almost anything can be arranged somewhere in North America, from pony treks and scuba trips to guided rock climbing ascents and off-piste skiing in the Rocky Mountains. If you can make your own way to the start point, prices will come down; if you can't, arranging group trips will be cheaper for solo travellers and couples than shelling out for bespoke outings.

• **Shop around** – small, local companies offer similar trips to the big boys for less (they're often better guides, too).

• **Be your own tour guide** – some driving routes (stand up, Alberta's Icefields Parkway) offer stunning nature tours for free. Why pay for someone else to drive you?

• **Follow the gasoline rule** – don't forget that if you're putting in the effort on foot, by pedal or by paddle, it will probably cost you less than if there's a motor vehicle involved.

Above: Canyon de Chelly National Monument is a smart alternative to the Grand Canyon.

HIGHLIGHTS

CLIMB SQUAMISH, BRITISH COLUMBIA, CANADA
Thin cracks snake up the perpendicular rock walls of Stawamus Chief in Squamish, offering everything from bouldering to some truly spectacular big-wall climbing for those who aren't squeamish about being perched half a mile above British Columbia. *p203*

RUTA PUUC RUINS, YUCATÁN, MEXICO
Maya ruins spill out of the forest all over Yucatán, and many are free to explore once you leave the tourist circuit. The Ruta Puuc takes in a string of less-visited Maya sites crowned by Chaac (god of rain) masks and geometric latticework. *p207*.

KOKE'E STATE PARK, HAWAI'I, USA
The state parks of Hawai'i are something else – as you might expect from an elemental landscape forged by volcanoes. Kaua'i's Koke'e State Park is one of the best, crossed by some 80 kilometres of outstanding hiking trails, taking in perilous clifftop lookouts and inland forests. *p208*

KAYAK THE SAN JUAN ISLANDS, WASHINGTON, USA
OK, you'll pay to hire a kayak to explore the calm, still waters of the Salish Sea off the San Juan Islands, but if you spot a whale or orca breach nearby, we promise you won't mind the expenditure. *p209*.

NORTH AMERICA

VANCOUVER & BRITISH COLUMBIA, CANADA

Compared to British Columbia, other outdoorsy places feel almost indoors. Canada's backyard is a place of wild beauty and unrestrained nature – follow the lead of locals and you can tap into nature in a big way on a small budget.

Maplewood Flats Conservation Area, North Vancouver

Managed by the Wild Bird Trust of BC, the Maplewood Flats Conservation Area is a tangle of trees, winding paths and protected wetland beaches that lure swallows, ospreys and bald eagles – and there are free guided nature walks the second Saturday of every month. *wildbirdtrust.org; free.*

Do the Grouse Grind, North Vancouver

You don't get the experience completely free (there's a C$15 fee for the gondola ride back to town) but the Grouse Grind hike up the side of Grouse Mountain offers spectacular views of downtown glittering in the water below, which is twice as thrilling as the touristy activities on top. *grousemountain.com; 6400 Nancy Greene Way; gondola fee C$15.*

Mountain-bike the North Shore, North Vancouver

Some 70km of challenging trails bust through mountain forest terrain on the steep slopes leading down to Vancouver's northern suburbs. Rent a bike from Endless Biking (endlessbiking.com; 1467 Crown St, North Vancouver; 4hr rental C$35) and get info on trails and access from the North Shore Mountain Bike Association

(nsmba.ca). *Free, once you have wheels.*

Bare all at Wreck Beach, Vancouver

Vancouver's citizens are an uninhibited bunch, and nowhere more so than at the city's favourite (and North America's largest) nudist beach. Follow Trail 6 into the woods and down the steps to find a friendly crew of sunburned regulars. *wreckbeach.org; via Trail 6, University of British Columbia; free.*

Lynn Canyon Park, North Vancouver

Amid a dense cluster of century-old trees, the main lure of this popular park is its suspension bridge (a free alternative to the famous span at Capilano), which sways over the river that tumbles 50m

Below: the suspension bridge in Lynn Canyon Park in North Van is an alternative to the better-known Capilano bridge. Right: Squamish in British Columbia has some superb rock climbing routes.

© Firefly Images | 500px

NORTH AMERICA

below. Hiking trails, swimming areas and picnic spots will keep you busy either side of the crossing. *lynncanyon.ca; Park Rd, North Vancouver; free.*

🖐 Strike gold at Goldstream Park, Vancouver Island

This swath of temperate rainforest, a 20-minute drive from BC capital Victoria, squeezes a lot of nature into its 4.5 sq km. The gold ran out in the 1860s, so today the park is famous for sightings of bald eagles, particularly during the annual autumn salmon run. A 700m trail leads to Niagara Falls, narrower but only 4m shorter than its famous Ontario namesake. *goldstreampark.com; Trans-Canada Hwy, Vancouver Island; free.*

🖐 See totem poles at Thunderbird Park, Victoria, Vancouver Island

The admission fee is worth paying to see the excellent Royal BC Museum (royalbcmuseum.bc.ca; 675 Belleville St; entry C$17), but there's no charge to wander round adjacent Thunderbird Park, where a line of weathered, genuine totem poles leads to two of the province's oldest buildings: the 1844 St Ann's Schoolhouse and 1852 Helmcken House, both from the fur-trade era. *Belleville St, Victoria; free.*

🖐 Hug a Douglas fir, Vancouver Island

The spiritual home of tree-huggers, Cathedral Grove is the place to (with help from your friends) throw a woody embrace around some of BC's oldest trees, including centuries-old Douglas firs more than 3m in diameter. Located between Parksville and Port Alberni, the forest is accessible via trails leading off the road through a dense canopy of vegetation. *bcparks.ca/explore/parkpgs/macmillan; MacMillan Provincial Park, Vancouver Island; free.*

🖐 Climb the Chief at Squamish

Bring ropes, harnesses, boots and protection to Squamish and test your grip against some of Canada's best bolted and trad routes. Stawamus Chief, the 700m granite wall just outside town, offers muscley, big-wall moves on sheer granite, following vertical cracks up the cliff. Local operators can kit you out with gear if you forget something. *Squamish; free.*

🖐 Wild swim in Lillooet

Every Canadian lake looks clean enough to dive into, for the simple reason that they are. Swimming at Seton Lake near Lillooet is like bathing in mineral water, 26 sq km of it. This scenic expanse of water was created by a dam on the Seton River and its still waters run to 460m deep – perfect for refreshing, sometimes chilly swimming. *bchydro.com; Lillooet; free.*

🖐 Follow the Pacific Marine Circle Route

This dramatic loops clocks up 263km as it circles around Vancouver Island, starting and ending in Victoria. En route, there are stops along rugged stretches of coastline, and views across Juan de Fuca Strait and the Olympic Mountains. It's great on four wheels, even better on two (hire bikes from US$30 per day). *vancouverisland.travel/road-trips/pacific-marine-circle-route; free.*

WINTER SPORTS WITHOUT THE STING

Every skier and snowboarder knows that Whistler Blackcomb is the slope to slide, but costs can mount up faster than the fine powder that dusts the flanks of 2181m Whistler Mountain. For winter thrills your bank manager would approve of, try the small, community-run ski area at Summit Lake (skisummitlake.com) near Nakusp, where day passes cost C$42, compared to C$128 at Whistler.

NORTH AMERICA

BEST
NATIONAL-PARK
WONDERS

The national parks are North America's wild playground. Some charge an entry fee, but once you're in, they're full of fantastic free escapades.

PYRAMIDS TALL AS TREES AT TIKAL NATIONAL PARK, GUATEMALA

The famous Maya pyramid temples at Tikal are just a tiny part of this 575 sq km rainforest reserve that throngs with toucans, monkeys and coatimundis, part of the 10,000-sq-km Maya Biosphere Reserve created to protect the Peten jungle. *Entry 150GTQ.*

MOON-WALKING, NEVADA, USA

Two hours' drive from the neon glow of Vegas, Nevada's night sky is so clear you can regularly see five planets. Rangers offer free lunar-lit night hikes in summer when there's a full moon. *nps.gov/ grba; Great Basin NP, Nevada; free.*

VOLCANO VOYEURISM, HAWAI'I, USA

Surf's up in Hawai'i, and so is the lava. Kilauea volcano remains active, and you can look out over the aftermath of the 2018 eruption from the Kilauea Overlook, including the dramatically remodelled Halema'uma'u crater. *nps.gov/havo; Hawai'i Volcanoes NP, Hawai'i; US$15 individual, US$30 car, 7-day pass.*

VERTICAL HIKING, MAINE, USA

The 3.2km Precipice Trail dramatically ascends the east face of Champlain Mountain, along super-narrow ledges with vertical climbs up vie ferrate-style rungs. *nps.gov/acad; Mount Desert Island, Acadia NP, Maine; opening hours vary; US$15 individual, US$30 car, valid 7 days.*

WILD HOT-TUBBING, USA

Park on the 45th parallel and follow the steam to Boiling River, a bathing spot where hot springs enter Gardner River to create perfect soaking conditions. *nps.gov/ yell; Mammoth Hot Springs, Yellowstone NP, Idaho, Montana & Wyoming; US$20 individual, US$35 car, 7-day pass.*

FLOWER POWER, USA

More than 1500 wild-flower species bloom beneath the Appalachians in the Great Smoky Mountains during spring. Go backcountry hiking and forest camping (hiking permits from US$4 per night; camping per night US$14 to US$23) amid the ephemerals. Watch out for bears! *nps.gov/grsm; Great Smoky Mountains NP, North Carolina & Tennessee; free.*

SNORKEL AROUND A FORT, FLORIDA, USA

Some 110km from Key West, perched on a coral atoll, Fort Jefferson is the scene of much marine activity. Snorkel around the moat and sea wall amid fish, coral and wrecks. *nps.gov/drto; Dry Tortugas NP, Florida; US$15 7-day pass.*

FIND BEARS, ALASKA, USA

Few encounters are as exciting as running into a big brown bear. Fortunately, at Brooks Camp, the bears have tastier fish to fry than you. July to September is the best time to see them. *nps.gov/katm; 24hr; Brooks Camp, Katmai NP, Alaska; free; camping US$12 per person.*

TRAVEL THROUGH SPACE AND TIME, NEW MEXICO, USA

A dark-sky park, Chaco has terrestrial wonders to complement its heavenly delights, with 1000-year-old Puebloan ruins, built to align with the stars during the equinox. Rangers offer free tours. *nps.gov/chcu; Chaco Culture National Historical Park, New Mexico; US$15 individual, US$25 car, 7-day pass.*

ROAM GROS MORNE'S FJORDLAND, CANADA

Worth going off the beaten track for, Newfoundland's World Heritage-listed Gros Morne National Park is an elemental landscape of fjords, inlets and coastal mountains, best explored on foot or by sea kayak. *pc.gc.ca/en/pn-np/nl/grosmorne; Gros Morne National Park, Newfoundland; May-Oct; entry US$10.*

RUTA MAYA, MEXICO & BELIZE

The former heartland of the Maya empire is a sprawl of beaches, ruins and jungles, making it a perfect place for budget Indiana Joneses. You can scramble over ancient ruins, swim in sinkholes, snorkel reefs and still have change for a beachside beer at sunset,

✋ Tulum Beach, Mexico

While you'll pay to explore the moody ruins at Tulum (M$75), the beach is free for all – a lovely expanse of white sand, between rocky headlands crowned by Maya pyramids and swaying palms. Access the sands via paths leading down between beach cafes, resorts and the Tulum archaeological site. *Tulum, Yucatán, Mexico. Free.*

✋ Reserva de la Biosfera Sian Ka'an, Mexico

Easily accessible from Tulum by rented bike or scooter, Sian Ka'an (Where the Sky is Born) has spider and howler monkeys, crocodiles, tapirs, giant land crabs, manatees and more than 330 bird species. The entry fee is small for so much wildlife, and wild camping is possible on the shore. *Sian Ka'an Biosphere Reserve, Yucatán, Mexico; entry M$36.*

✋ Hormiguero, Mexico

This rarely-visited Maya city (whose modern name is Spanish for 'anthill') flourished during the late Classic period, and getting here is an adventure on an overgrown, pothole-riddled road. On arrival, you can explore the castle-like ruins of Estructura II, with its monster-mouth doorway and ornately decorated towers. *22km southwest of Xpujil, Mexico; free.*

✋ El Volcán de los Murciélagos, Xpujil, Mexico

Every evening at this cavern, inland from the coast near Xpujil, some two to three million bats swirl up from the depths of a dry cenote (sinkhole), forming a tornado of fur and wings that's a surreal experience. Look for the bat crossing signs on Hwy 186, about 500m from the turnoff. *Zona Sujeta a Conservación Ecológica Balam Ku; Xpujil, Mexico.*

✋ Cenote Azul, Laguna Bacalar, Mexico

The on-site bar and restaurant brings a touch of civilisation at this cenote, 3km south of Bacalar town centre, but the eerily dark waters (90m at the deepest point) are fringed by forest and are a

Below: Mayan ruins with a view (and a beach) at Tulum; they're more commercial these days but smaller Mayan sites are not far away. Right: the cenote at Laguna Bacalar.

© Florian Trojer | Getty Images

gorgeous spot for swimming. You can get here easily by highway bus, and the entry is a small cost for such a pretty spot. *Hwy 307 Km 34, Laguna Balacar, Mexico; entry M$25.*

✋ Pedal to cenotes at Cobá, Mexico

A modest M$50 will get you a rental bike for the day at Cobá, providing access to a string of gorgeous swimmable cenotes secreted away in the forest at Choo-Ha, Tamcach-Ha and Multún-Ha. There's an entry fee for each, but these partly-hidden waterholes are blissful places to dive into crystal waters. *Cobá, Mexico; cenote entry M$100.*

✋ Kinich-Kakmó Pyramid, Izamal, Mexico

Three of Izamal's 12 Maya pyramids have been partially restored and, unlike at Chichén Itzá, there's no fee to clamber up steps which may once have hosted human sacrifices. Three blocks north of the town's monastery stands 34m-high Kinich-Kakmó where, legend has it, a deity in the form of a blazing macaw would swoop down from the heavens to collect offerings. *C/27, Izamal, Mexico; free.*

✋ Ruta Puuc Ruins, Yucatán, Mexico

Inland from the coast, the Ruta Puuc is dotted with less-visited Maya ruins. Make for the ornate palacio at Xlapak; it's one of the most elegant structures on the Ruta Puuc, and it's free to explore. The palace is crowned by a headdress of Chaac masks, columns and fretted geometric latticework. *Ruta Puuc, Mexico; free.*

✋ Punta Esmeralda, Playa del Carmen, Mexico

Finding a quiet stretch of sand at Playa del Carmen can be tricky, but

Emerald Point is a favourite retreat for knowledgeable locals, set on the northern edge of the city, where a shallow cenote creates a natural pool framed by a soft white-sand beach. It's an idyllic spot to spread your beach towel for free. *Off 5 Av Norte, Playa del Carmen, Mexico; free.*

✋ El Castillo Real, Isla Cozumel, Mexico

On Isla Cozumel, the rough, pitted road to Punta Molas passes the sprawling Maya ruins known as El Castillo Real (The Royal Castle). The archaeological site is heavily eroded, but there's a pleasing wildness to this little-visited site on its lonely, wave-lashed shore. The bumpy cycle here is part of the fun. *Puntas Molas, Cozumel, Yucatán, Mexico; free.*

✋ Lamanai, Belize

One of Belize's most fascinating Maya sites lies 58km south of Orange Walk Town on an unpaved road (or a 38km trip up the New River). An acceptable fee grants access to a marvellous setting, surrounded by dense jungle overlooking the New River Lagoon. Climbing to the top of the 38m High Temple to gaze out across the vast jungle canopy is an awe-inspiring experience. *Orange Walk District, Belize; entry BZ$10.*

FREE SNORKELLING IN YUCATÁN

The best snorkelling sites on the Yucatán coast have entry fees and mandatory guides; to snorkel for free, bring your own gear and splash in off the beach. A resort lunch will get you access to Tankah Bay, with splendid, fish-crowded sites off the northern end of the beach; or there's good off-beach snorkelling at Playa Garrafón and Yunque Reef on Isla Mujeres. To go off-piste, scramble along the shoreline north of Tulum Beach, where cuttlefish are common along the rocky shore.

NORTH AMERICA

HAWAI'I, USA

Hawai'i has a reputation for money-draining experiences and lavish living, but there are many ways to save, from free nature walks to basking on wild beaches and surfing gnarly breaks on the islands where the sport was invented.

🤚 Beach life

There's no charge to use most of Hawai'i's blissful beaches, though you'll burn through the bucks staying near famous sands such as Waikiki. Seek out quieter, more secluded stretches, such as Mākua Beach on O'ahu and Kawakiu in Moloka'i's wild west, where there are fewer financial distractions. *Across Hawai'i; free.*

🤚 Niaulani Nature Walk, Hawai'i Volcanoes National Park, Big Island/ Hawai'i

On Mondays, free guided nature walks explore Volcanoes NP's rainforest, with guides explaining the ecological importance of old-growth koa and ohia forests, traditional uses of plants, and the role of birds in the ecosystem. *volcanoartcenter.org; Volcano Art Center, 19-4074 Old Volcano Rd, Big Island/ Hawai'i; free.*

🤚 Kahuku Unit, Ka'u, Big Island/ Hawai'i

Kahuku Unit's six hiking trails lead through green pastures to volcanic cinder cones, lava tree-molds, rainforests and lava flows. Hikers can wander freely Wednesday to Sunday, but the experience comes alive during excellent ranger-guided hikes on Sundays. *nps.gov/havo/planyourvisit/kahuku.htm;*

off Hwy 11 at Mile 70.5, Ka'u, Big Island/ Hawai'i; free.

🤚 Lawai International Center, Kaua'i

Open for tours that feel like a mini pilgrimage, this sublimely peaceful site once held a Hawai'ian *heiau* (temple), but Japanese immigrants placed 88 miniature Shingon Buddhist shrines along a steep hillside path, symbolizing the pilgrimage shrines of Shikoku. *lawaicenter.org; 3381 Wawae Rd, Lawai, Kaua'i; by donation.*

🤚 Koke'e State Park, Kaua'i

Koke'e State Park is the starting point for almost 80km of outstanding hiking trails to perilous clifftop eyries, to the woods overlooking Waimea Canyon, and to Alaka'i Swamp, which teems with native bird species. *hawaiistateparks.org/kauai; Koke'e & Waimea Canyon, Kaua'i; free, parking US$5.*

🤚 Pu'u 'Ualaka'a State Wayside Park, O'ahu

For the best free vista in Honolulu head to this lofty hillside park. Sweeping views extend from Diamond Head on the left, across Waikiki and downtown Honolulu, to the Wai'anae Range on the right, with the gleaming Pacific as a backdrop. *hawaiistateparks.org; 2760 Round Top Dr, Honolulu, O'ahu; free.*

PACIFIC NORTHWEST, USA

The Pacific coast cuts a rugged line from Northern California to the Canadian border, taking in the wildest country in the west. There are big distances to cover, but also big adventures, on forested mountains and in whale-filled oceans, for pocket prices.

✋ Forest Park, Portland, Oregon

To get away from it all without leaving the city, take a walk in the country's largest urban park. Decked in moss, ferns and other primordial flora, the paths wind through towering trees – crick your neck in wonder as you gaze up to their dizzying heights. *www.forestparkconservancy.org; Northwest Portland; free.*

🚗 Drive the North Cascades Hwy, Washington

Scenic drives are what the Pacific Northwest does best, and the North Cascades Hwy – aka State Route 20 – is the top of the heap. Like driving into a pioneer's photograph, the road strains past emerald forests, sawtooth mountains, hiking trails and milky blue lakes that once lured Gold Rush prospectors. *Washington; free.*

🐋 Whale-watching at Mendocino Headlands State Park, California

Mendocino is famous for its whale festivals, when folk gather for guided whale-watching walks for just US$5, but you can see whales almost anytime from the cliff paths that cross the rocky headlands. The Point Arena Lighthouse is a great place to scan the horizon for flukes. *parks.ca.gov; Mendocino, California; free.*

🛶 Kayak the San Juan Islands, Washington

The cost of renting a kayak in the San Juan Islands is insignificant when you factor in the chances of encounters with cetaceans. Paddle along pine-fringed coastlines to rocky islets where you may spot surfacing porpoises and whales. Try San Juan Island Outfitters (sanjuanislandoutfitters.com; Friday Harbour; kayak hire per hour US$25).

🐋 Orcas at Lime Kiln Point, Washington

Orcas are the stars at Lime Kiln Point on San Juan Island, though whales and porpoises also appear. Pick a clear day, stake out a point on the cliffs and watch the show. *parks.state.wa.us/540/Lime-Kiln-Point; San Juan Islands, Washington; free.*

Below: watch for (free) whales off the Californian coast at Mendocino.

NORTH AMERICA

COSTA RICA

Rainforests envelop Central America, and Costa Rica is where they erupt into their most extravagant finery. Bring a sense of wonder, not a thick wallet, for an audience with a sloth, a tête-à-tête with a volcano, or a date with a surf break.

✋ Surf the Nicoya Peninsula

Barefoot rules apply on the Nicoya Peninsula, where Costa Rica's best-loved breaks cut left and right along a shoreline dusted with golden sand. The laid-back hippy scene is cheap to plug into – budget camping, hostels and cabanas, cheap board hire, shorefront yoga, and beers on the beach. Start at Mal País and Santa Teresa. *Board hire from US$15.*

✋ Hike to Montezuma Waterfalls

The most popular dry-land activity drawing surfers from the Montezuma spray is this hour-long hike into the forest, to a waterfall with a delightful swimming hole. The main thrill is leaping 12m into the water at the upper falls – not a splash for the faint hearted. Parking at the start of the trail is US$2. *Montezuma; free.*

✋ Parque Nacional Cahuita

This pocket-sized national park covers just 10 sq km but takes in white-sand beaches, coral reefs, rainforest and a lagoon bursting with wildlife, including armadillos, sloths and howler monkeys. Enter from Kelly Creek and you don't need to pay the US$5 entry fee charged at the Puerto Vargas entrance. *Cahuita; US$5, free from Kelly Creek.*

✋ Find a free hot spring at La Fortuna

Resorts have snaffled up the prime hot-spring bathing spots at La Fortuna, but any local can point you towards free places to sample the waters – ask about El Chollín, a thermal river that creates a surging hot-tub that's the antithesis of the sanitised spa experience. *La Fortuna de San Carlos; free.*

✋ Mountain-bike down a lava flow

Mountain-biking is erupting in a big way on the Arenal volcano. The company Arenal 1968 (arenal1968.com; El Castillo–La Fortuna road; trails US$15) has a web of trails crossing the original 1968 lava flows; rent wheels from Bike Arenal in La Fortuna (bikearenal.com; cnr Ruta 702 & Av 319A, La Fortuna de San Carlos; bike hire per day UDS$15).

Below: don't pay resort prices to soak in hot springs near Arenal volcano.

© Ryan Henke | 500px

THE FOUR CORNERS, USA

The meeting point of Arizona, Colorado, New Mexico and Utah is the living image of the Wild West: burnt bronze deserts, frontier outposts, Native American nations and national monuments carved by nature – with all sorts of free thrills for modern-day pioneers.

✋ Scenic driving, Moab, Utah

Moab's legendary national parks – Arches and Canyonlands – charge hefty entry fees. For free geology, point your car towards the Upper Colorado River Scenic Byway U-128, where you can gaze down on the Colorado River. *discovermoab.com/scenic-byway-u-128; Moab, Utah; free.*

✋ Grand Staircase-Escalante National Monument, Utah

Despite efforts to cut its size, this vast, free wilderness protects 6879 sq km of desert terrain. Infrastructure is minimal, leaving a huge area of arches, outcrops and canyons for adventurers; Kanab visitor centre has maps and advice. *blm.gov/visit/kanab-visitor-center; 745 Hwy 89, Kanab, Utah; free*

✋ Garden of the Gods, Colorado

It was written into law that the Garden of the Gods – an otherworldly rocky outcrop northwest of Colorado Springs – should be free to all. Locals go climbing, hiking and picnicking around the spires. *gardenofgods.com; Colorado Springs, Colorado.*

✋ Ski for free, Colorado

It's the skipass that costs money, so look for ski areas offering free 'Uphill Access' passes, where you can hike, snow-shoe, or ski up (using grippy 'skins') and then whoosh downhill cost-free. Try Aspen (aspensnowmass.com), Vail (vail.com), and Copper Mountain (coppercolorado.com).

✋ Find a vortex at Sedona, Arizona

With a US$5 Red Rock day pass, you can hike into National Forest land to amazing mesas (plateaus) that locals believe are crossed by vortexes, swirling energies that channels the earth's power. *visitsedona.com; Sedona, Arizona; day pass US$5.*

✋ Go wild in the badlands, New Mexico

You're off the map in the Bisti/De-Na-Zin Wilderness, 180 sq km of badlands near Farmington, New Mexico. Bring a GPS to roam this alien landscape of hoodoos (eroded spires). *blm.gov/visit/bisti-de-na-zin-wilderness; Farmington, New Mexico; free.*

Below: hike Coyote Gulch in the Grand Staircase-Escalante desert and stay overnight under the stars.

NORTH AMERICA

WELLNESS

Wellness is big, expensive business in the US, but the bill comes down as you head south into Central America. For spas in particular, there's a tangible shift from luxury to laid-back as you cross the US border and drift south along the beaches of Mexico, Guatemala and beyond.

Across the region, the cheapest wellness experiences are provided by nature: soaking in natural hot springs in California's national forests; swimming in jungle pools on the Ruta Maya; aurora-spotting in Alaska; the polar bear swim on New Year's Day at English Beach in Vancouver.

Above: experiencing awe under the aurora borealis in Fairbanks, Alaska. Right: strike a yoga pose in Oregon in the Pacific Northwest.

Organised wellness at the yoga and healing end of the spectrum tends to coincide with an outdoor way of life: it's big in the Pacific Northwest, across California, in the beach-focused Caribbean and south along the coast of Yucatán in Mexico.

YOGA, MEDITATION AND SPIRITUALITY

California is the USA's capital of wellness – along with everything else holistic and new-agey – but complementary therapies aren't always complimentary. You're more likely to find classes by donation than completely free yoga lessons, for example.

In the Caribbean and Central America, wellness is a beach-based activity, with morning yoga sessions on the sand, and massages under pavilions looking out over the waves. Hotel spas target the top end of the market, but there are also small, independent centres, not affiliated to big hotels, where even thrifty travellers will find affordable relaxation.

• **Yoga with lunch?** – some of the most affordable yoga classes are offered by traveller restaurants alongside healthy, often vegetarian or vegan, food.
• **Check the board** – noticeboards

in traveller centres and student neighbourhoods are the place to find inexpensive tuition and drop-in classes.
• **Buy in bulk** – if you're sticking around in one spot, buy a block of classes and pay less than individual drop-in rates.
• **Go it alone** – pack a yoga mat and every beach, viewpoint or natural spring becomes a place for stretching or contemplation.

NATURAL WELLNESS

Nature throws in a lot of wellness for nothing in North America, and the continent had its own traditions of healing long before Europeans came to these shores, based on herbal medicine, spirituality and harmony with the natural environment. Lingering hints of these traditions endure in Native American and First Nations communities.

Modern North Americans still make full use of the hot springs that bubble up all along the west coast, from the budget-friendly waters at Umpqua in Oregon to the naturist-friendly wild springs at Deep Creek in San Bernardino National Forest and the Hierve el Agua in Mexico's Oaxaca.

• **Get into hot water** – even at developed hot-spring resorts, locals can usually point you towards free, natural pools they use themselves, away from the organised spas.
• **Swim in nature** – swimming in waterfalls, streams and lakes surrounded by nature awakens a primal sense of wellbeing; keep a towel handy for the next glacial lake or forest cascade (you might not even need a costume).
• **Forage the forest** – wild food bursts from the landscape in Canada and the US; look for blueberries in the mountainous north, and wild strawberries and edible mushrooms everywhere.

HIGHLIGHTS

TECOPA MUD BATHS, CALIFORNIA

For a spa-style freebie, smear yourself with mineral mud at this secret California hot spring, then bake to a crust in the Death Valley sunshine, and wash off in balmy, geothermally heated waters, surrounded by an empty desert landscape. *p214*

DO YOGA ON A PADDLE-BOARD IN THE CAYMAN ISLANDS

Some advanced yoga poses can be tricky enough on solid ground; floating on a stand-up paddle-board in a turquoise lagoon adds a whole new dimension. Vitamin Sea runs classes all around Grand Cayman in which you can combine these beachy disciplines. *p216.*

NORTH AMERICA

CALIFORNIA, USA

California caught the wellness bug during the Summer of Love and never shook it off. From enthusiastically bendy yoga classes in San Francisco to mud-packs at only-known-to-locals hot springs near Death Valley, wellness doesn't cost an arm and a leg.

🖐 Budget-friendly Bay Area yoga

Free love may have moved on, but free yoga is still a thing in SF...well, almost. Numerous studios offer yoga classes by donation, with a share of proceeds going to local charities. Try Laughing Lotus in Mission Dolores (sf.laughinglotus.com; 3271 16th St) or Oakland's Flying Studios (flying-studios.com; branches at 4308 & 4834 Telegraph Ave). *Recommended donation from US$10.*

🖐 Take a mud bath at Tecopa

California spas offer lavish treatments for lavish prices, but at Tecopa, you get the spa vibe for free. Slather yourself with mineral mud at the natural hot springs just north of town, bake dry in the Death Valley sunshine, then scrub clean in the balmy, 40°C waters. It's a casual, classic California experience that will introduce you to some interesting folks. *Tecopa, Inyo Country; free.*

🖐 Vegan feasts at Millennium, Oakland

Oaklanders rejoiced when San Francisco's classiest vegan restaurant hopped across the bay in 2015, bringing treats such as panko-crusted maitake mushrooms along with it. The creative plant, fungus and grain-based dishes are made with mostly organic, locally grown ingredients. *millenniumrestaurant.com; 5912 College Ave, Oakland; mains from US$15.*

🖐 Hot springs in a forest glade

Some of California's best hot springs are in Los Padres National Forest, inland from Santa Barbara. Trek into the woods to find secluded – and free – natural hot-spots such as Sespe and Sykes, where the only thing to disturb your natural soak might be a wandering bear. *www.fs.usda.gov/lpnf; US$5 day pass to reach some areas.*

🖐 Go all in at Scarlet Sage, San Francisco

Do you have a spiritual need involving herbal remedies, natural therapies, essential oils or crystals? This store offers it all, plus classes in emotional freedom technique, tarot card reading, womb healing, astrology – the full California package. *scarletsage. com; 1193 Valencia St, San Francisco.*

Below: Scarlet Sage's range of wellbeing products.

NORTH AMERICA

SEATTLE, USA

Seattle has a history of doing things its own way – in other words, perfect conditions for innovative takes on wellness. Not everything is free, but nearby nature provides plenty of ways to shake off city life and exhale.

✋ Alternative takes on yoga

Never one to the follow the crowd, Seattle is putting all sorts of novel spins on the traditional yoga class. For standard yoga-class prices you can try 'noise yoga' – stretching to industrial-sounding, experimental hums, buzzes and drones, or even metal yoga, to a soundtrack of thrashing guitars. Check Seattle message boards for classes. *Prices vary.*

✋ Find Flora in Seattle

A longtime favourite for Seattle vegans and veggies, Flora has a garden-like vibe and a lively, creative menu. Dinner treats include fried avocado, yam fries, grilled asparagus pizza and black-bean burgers; or go for the hoppin' John fritters or tomato-asparagus scrambles at brunch. *cafeflora.com; 2901 E Madison St, Madison Park, mains from US$15.*

✋ Swim al fresco

Seattle residents aren't put off by the sometimes chilly (sometimes icy) waters. Swimming in the open water is possible all round Seattle, from freshwater lakes to the waters of Puget Sound, and it's undeniably invigorating. Take your first freshwater splash at Matthews Beach Park or Green Lake Park, or hit the salt water at Seahurst Park. *Free.*

✋ Feel the presence of the whales

Seeing these magnificent creatures breach from a vantage point on the shore is always an awesome experience. You don't need to roam far out of Seattle: Point Defiance, Point No Point, and Point Robinson on Maury Island are all good spotting sites. *Free.*

✋ Find yourself in the forest

Head out into the forests around Seattle and you'll discover how the Evergreen State got its name. To be alone amongst the moss-draped trees is to strip away the trappings of modern life and rediscover what it means to be alive. Start the voyage of discovery inland from Seattle at emerald-green Mount Baker-Snoqualmie National Forest. *fs.usda.gov/mbs; day-use free.*

Below: take time to follow forest-bathing practices when hiking in the Mount Baker Snoqualmie National Forest.

NORTH AMERICA

THE CARIBBEAN

The scattered islands of the Caribbean are like cut emeralds set in a turquoise tiara. It's no surprise that lush, luxury spas have sprung up all over these idyllic isles; it's more surprising to find wellness experiences that don't need an expense account.

Enjoy a natural whirlpool in the Virgin Islands

Beachside rock outcrops create a vigorous natural whirlpool at Bubbly Pool, north of Little Harbour in the British Virgin Islands. When the waves crash into this narrow inlet, the surging water fills a pool with bubbles creating a natural saltwater Jacuzzi. *Jost Van Dyke, British Virgin Islands.*

Do yoga on a paddleboard in the Cayman Islands

Love yoga? Adore stand-up paddleboarding? Why not combine the two? For less than the cost of a leisure boat trip, you can experience floating yoga classes over turquoise water at beaches around Grand Cayman. *vitaminseacayman.com; 10 Market St 269, Georgetown, Grand Cayman; classes US$25.*

Take a Cuban health cure

Cuba's nostalgic Balneario San Diego bathhouse feels more like a 1950s hospital than a luxury spa, but in place of fluffy towels and fancy facials, you get restorative baths in sulphurous thermal waters (30°-40°C). *San Diego de los Baños, Cuba; baths CUC$8.*

Cold cascades in the Dominican Republic

There's a tiny fee and a sweaty 7km hike to reach Salto de Jimenoa Uno, but the sight of the waters plunging 60m over a rocky cliff make it all worth it. Swimming here is icy cold, but highly invigorating. *Jarabacoa, Dominican Republic; entry RD$100, including a bottle of water.*

Swim from island to island

For fit swimmers, there are abundant opportunities for inter-island swimming, from short, exhilarating swims to islets offshore to serious distance swims that need boat back-up. For (paid-for) supported swims, consider the races from Nevis to St Kitts (nevistostkittscrosschannelswim. com) and the St Croix Coral Reef Swim (swimrace.com).

Bathe in bioluminescence

There's a chance you'll stumble across bioluminescence spontaneously, but to enjoy this almost supernatural experience on demand, head to Jamaica's Glistening Waters, where a US$25 fee gets you access to a lagoon lit up like a trance club by glowing-green algae. *glisteningwaters.com; Falmouth, Jamaica; boat trip US$25.*

NORTH AMERICA

YUCATÁN & QUINTANA ROO, MEXICO

The states of Yucatán and Quintana Roo sit pretty on Mexico's Caribbean coast, sandwiched between ruins-filled jungles and icing-sugar beaches. Even on a budget, it's easy to plug into local Maya traditions of wellness and unspoiled nature.

Oceanside yoga on Isla Mujeres

Head to the Treehouse at the ever-popular Buho's Bar for inexpensive yoga classes right by the water, in a sand-floored garden of swishing palms. Afterwards, pull up a swing (it's what they have in place of barstools) at the bar for a cold cerveza. *facebook.com/YogaAtTheTreehouse; Buho's, C/Carlos Lazo 1, Playa Norte, Isla Mujeres; yoga from US$10.*

Spring-bathing at Síijil Noh Há, Riviera Maya

Síijil Noh Há is run by the local Maya community; for a small entry fee, you can bask in a spring-filled cenote (unusually, wide open to the sun's rays) set on a gorgeous lagoon, or go kayaking and hiking, before sampling authentic Maya food. *facebook.com/siijilnohha; off Hwy 307, Laguna Ocom turnoff, Felipe Carrillo Puerto, Riviera Maya; entry M$20.*

Experience Maya medicine

The traditional medicine of the Yucatec Maya is a whole-body system of healing, fusing herbalism, cleansing therapies and spirituality. Consultations and treatments are available at the Jardín Botánico Medicina Herbolaria Yaxcabá, one of the last remaining centres. *facebook.com/jardinbotanico.yaxcaba; C/22, Yaxcabá; costs vary.*

Go to a barefoot gallery, Tulum

SFER IK at the Azulik Resort is a surreal space of polished concrete waves and windows woven from bejuco (vine-like wood) that look onto the forest. Visitors are invited to walk barefoot to experience the site as a living organism. *sferik.art; Carretera Tulum-Punta Allen Km 5, Riviera Maya; free.*

Veganism with a view, Laguna Bacalar

After relaxing in the palm-framed waters of Laguna Bacalar, enjoy quality vegan food. Easy-going Mango y Chile has an airy deck overlooking the lagoon, where you can tuck into vegan tacos, salads and plant-based burgers. *facebook.com/mangoychile; Ave 3, btwn calles 22 & 24, Laguna Bacalar, Costa Maya; mains from M$100.*

Below: combine tropical beachlife with healthy vegan food at Laguna Bacalar.

NORTH AMERICA

THE BEST THINGS IN LIFE IN SOUTH AMERICA: TOOLKIT

With great variety and low-stress border crossings, South America was tailor-made for overland travel, but make your pennies go further by self-catering in local hostels, street-food grazing and soaking up bounteous nature outside the national parks.

Getting the best out of South America on a budget takes some planning. If you want to get deep into national parks and tick off bucket-list adventures like trekking in the Amazon and riding with gauchos in Argentina, your costs will rocket.

But venture off the tourist trail and it's possible to get by on less than US$50 per day, taking advantage of local buses, inexpensive hostels and cheap-and-cheerful local restaurants. Look for adventures you can take independently, using public transport, a bike, or a stout pair of hiking boots.

TOP BUDGET DESTINATIONS IN SOUTH AMERICA

Exploring this supersized continent can be expensive, so make your money count by making the most of modest costs in the following stops:
• **COLOMBIA:** The backpacker's favourite, full of jungle thrills and spills at pocket-pleasing prices.
• **BOLIVIA:** Epic landscapes, rich culture and good-value *hostales* and *almuerzos* (set lunches).
• **ARGENTINA:** Offset pricey city living

with cheap coastal stops and free hikes in the interior.
• **PERU:** Swap the mountain ruins for the coastal desert and your costs will dive.
• **PARAGUAY:** Waterfalls, river rides, national parks and colonial-era relics all come cheap.

TRANSPORT

Never underestimate the distances involved in South America. It's 7617km from Colombia's Point Gallinas to Cape Horn in Chile, with major obstacles en

Below: cycling in Buenos Aires, Argentina; many South American cities, especially in Colombia and Argentina, have good biking infrastructure.

© Philip Lee Harvey | Lonely Planet

route in the form of the Andes mountains, Amazon rainforest and Patagonian Steppe. Overland travel can be cheap but painfully slow, pushing many travellers to fly to avoid interminable, stomach-flipping drives on gravel roads.

AIR

Domestic and country-to-country flights in South America can be agreeably inexpensive, particularly in Bolivia, Ecuador and Peru. However, schedules are limited and airports are often far from city centres (taking a taxi to town will increase journey costs significantly).

AIR PASSES

OneWorld's Visit South America Airpass (oneworld.com) includes Argentina, Bolivia, Brazil, Chile, Colombia, Ecuador, Paraguay, Peru, Uruguay and Venezuela, plus Easter Island and the Galápagos Islands, offering maximum range for your dollar.

Consider local air passes, too. Brazil's Gol Airlines has a cost-effective South America Airpass (voegol.com.br) that covers Brazil, Argentina, Bolivia, Chile, Paraguay and Uruguay. Chile's LATAM Airlines (latam.com) offers a similar package. Also look out for single-country passes with Avianca (avianca.com), Azul Airlines (voeazul.com.br) and GOL (voegol.com.br).

• **Fly long, bus short** – save air travel for long-distance trips and take the bus for shorter hops: the planet, and your wallet, will thank you.
• **Fly like a local** – in Argentina, LADE (Líneas Aéreas del Estado; lade.com.ar) is a military-operated airline offering cheap seats to remote locations.
• **Avoid the holidays** – demand and prices soar around Christmas, Easter and holidays such as national day celebrations.

TRAIN

Most countries have railways, but services stop abruptly at international borders. Where you can find them, trains are cheaper than buses, but services are less frequent and some locos limp along at a snail's pace. Useful lines include: Trenes Argentinos (argentina.gob.ar/transporte/trenes-argentinos), trains from Buenos Aires; Empresa de los Ferrocarriles del Estado (efe.cl), trains from Santiago, Chile. For services from Cuzco and Lima, Peru Rail (perurail.com), Inca Rail (incarail.com) and Ferrocarril Central Andino (ferrocarrilcentral.com.pe).

BUS

Buses are the engine that drives South America, with services linking almost everywhere, in varying standards of reliability and comfort. In the Andes, many buses seem barely roadworthy (and many roads seem barely worthy of the name 'roads').
• Bus stations (known variously as *terminales de autobuses, terminales de ómnibus, rodoviária* and *terminal terrestre*) are often out of town – take a local bus to get to them.

Above: buses are an affordable way to get around, with vehicles ranging from local buses to long-distance coaches.

SOUTH AMERICA

• **Plan ahead** – book tickets a few days before travel, wherever possible; note that services slow to a trickle on religious holidays.

• **Travel by night** – *coche-cama* (sleeping car) buses offer reclining bed-seats for a premium; *lujo* (luxury buses) have seats that partially recline, offering a better chance of getting to sleep than standard seats.

BOAT

Long-distance boats travel on major rivers such as the Orinoco or Amazon. Prices are low, but so are comfort levels – unless you pay extra for a cabin, you'll spend nights in a hammock on deck (carry bug repellent). There are a few useful coastal ferries, including Chile's popular Navimag (navimag.com) linking Puerto Montt to Puerto Natales.

CAR

Many hire companies only rent to over 25s; rates range from US$40 to $80 per day, with big surcharges if you don't return the car to the same location. Note that many hire companies don't allow their vehicles to be taken over international borders.

• **Bring the right papers** – most hire companies ask for an International Driving Permit alongside your home license.

• **Don't scrimp on insurance** – South America's roads crunch up cars for breakfast, so check the conditions of the *seguro* (insurance) on any rental agreement.

• **Take safety seriously** – make sure you're fully insured, park in secure parking areas with a security guard, drive with locked doors, and never leave anything in a car overnight.

Above: from the Brazilian river port of Manaus, ferries depart from destinations along the Amazon, including all the way to Belém.

© LUC KOHNEN | Shutterstock

LOCAL TRANSPORT

Most big cities have reliable urban transport; usually local buses, occasionally trains and always taxis. Bus tickets are cheap. Taxis are often expensive, and kidnappings (for robbery or worse) are not unheard of; never get into one if there is someone inside other than the driver.

• **Uber me** – Rideshare apps are gaining a foothold in South America, but given the security risks, many prefer not to use them. Uber, Cabify and Easy Taxi are major players.

• **Meters** – are rare, and meters that are actually used are rarer: negotiate a fare before you start the trip.

• **Know your cabs** – for security, stick to official, government-licensed taxis with recognisable livery, or get your hotel to book you one.

• **Watch your stuff** – be wary of bag-snatchers and pickpockets on any form of public transport.

ACCOMMODATION

Accommodation prices in South America are a breath of fresh air compared to Europe or North America but, as everywhere, comfort costs. If you don't mind staying in hostel dorms, you can breeze through South America paying less than US$15 per night for accommodation.

CAMPING

Wild camping is commonplace in Argentina, tolerated in the countryside in Chile, Ecuador, Peru, Brazil and Ecuador, but not really recommended in Colombia or Bolivia for security reasons. Avoid camping near buildings or on private land (or ask the landowner first).

• **Bring all the gear** – avoid low-quality local rental gear: bring a tent, a stove and water bottles so you can purify water as you go.

• **Use designated sites** – in popular walking areas, official campgrounds offer hot showers, kitchens and a basic provisions store; rates start as low as US$5 per night.

• **Take refuge** – on some mountainous routes, *refugios* (basic huts with bunk spaces and kitchen access) are available, but prices can be high compared to sleeping under canvas.

HOSTELS, HOTELS & GUESTHOUSES

Albergues (hostels) are popular throughout South America. Official *albergues juveniles* (youth hostels) offer small discounts for Hostelling International/American Youth Hostel (HI-USA; hiusa.org) members.

The cheapest private rooms are at simple guesthouses, known as *hospedajes, casas de huéspedes, residenciales, alojamientos* or *pensiones*. Bathrooms are normally shared and hot water isn't always provided. Look out for cheap *casas familiares*, family houses where home cooking and hospitality make for excellent value.

• **Hotels** – actual *hoteles* (hotels) cost more, sometimes much more. Don't confuse hotels and hostels with *hostales*, which are small hotels and guesthouses in rural areas, serving up moderately priced rooms and filling meals for a largely local crowd.

• **Book ahead** – online booking agents have a good reach in South America.

• **Self-cater** – most hostels and some guesthouses have kitchens, meaning inexpensive meals if you don't mind skipping the tasty local cuisine.

• **To lodge or not to lodge** – the sky is the limit when it comes to prices for national park lodges; look for cheaper options just outside park boundaries and make day-trips.

ARTS & CULTURE

South American culture is a mesmerising blend of the old world and the new, of indigenous traditions and imports from Spain and Portugal, blended together with a sprinkle of revolution. It's a heady mix, and there's plenty that's free, though fees often crop up in unexpected places.

With religion at the region's heart and myriad social movements pushing power for the people, cities offer rich pickings for frugal culture vultures, with hundreds of free (or inexpensive) museums and galleries, historic churches and cathedrals, and grandiose monuments to revolutions.

Above: one of the world's most remarkable religious sites is Santuario de Las Lajas in southern Colombia. Right: the Chavin de Huántar ruins.

SOUTH AMERICA'S TOP SWITCHEROOS

With the added cost of getting to remote locations, aim to keep down costs when seeing South America's top sights and consider the following swaps.

- **SKIP:** Machu Picchu, Peru (entry US$45)
- **SEE:** Chavín de Huántar ruins, Peru (entry US$4.20)
- **SKIP:** Iguazu Falls, Brazil/Argentina (entry US$7.50-11.50)
- **SEE:** Catarata Yumbilla, Chachapoyas, Peru (entry free)
- **SKIP:** Statue of Christ the Redeemer, Rio de Janeiro, Brazil (entry US$14)
- **SEE:** Santuario de Las Lajas, Potosí, Colombia (entry free)
- **SKIP:** Sky Costanera, Gran Torre Santiago, Santiago, Chile (entry US$20)
- **SEE:** Parque das Ruínas viewpoint, Rio de Janeiro, Brazil (entry free)
- **SKIP:** Museu de Arte de São Paulo, São Paulo, Brazil (entry US$6.50)
- **SEE:** Museo Nacional de Bellas Artes, Santiago, Chile (entry free)

SOUTH AMERICA'S CULTURE FOR LESS

South America's top museums and galleries are often overlooked in favour of the Mets and Louvres of this world, but there are some spectacular collections, particularly in big cities such as Rio de Janeiro, Santiago and Lima. Reflecting revolutionary principles, many of the best are free to visit, or free one day a week to keep art accessible to the people.

- **Check the free days** – many big-city museums and galleries have no charge on Sundays, or on one Sunday a month.
- **Cities vs ancient sites** – the biggest museums are in capital cities, but some of the best are small museums associated with archaeological sites, often with low entry fees.
- **Show your ID** – many museums, galleries and monuments give student discounts; the International Student Identity Card (ISIC; isic.org) is recognised widely, particularly in Brazil, Peru and Argentina. Seniors also get decent discounts.
- **Guide fees** – historic sites can be free (or cheap) to visit, but you might be required to hire a guide, particularly at sites run by indigenous communities.
- **Think municipal** – many of South America's administrative buildings are grand, colonial-era institutions that are open to visitors for free with advance booking.
- **Let nature do the lifting** – as everywhere, tall buildings charge for access, but in mountain cities, there's almost always a natural viewpoint offering the same view for free.
- **A fee for nature?** – waterfalls, mountain viewpoints and other natural features are not always free; local communities may request admission or hiking fees.
- **Sacred art** – South America is awash with historic churches and cathedrals, open to all for free.

HIGHLIGHTS

MUSEO NACIONAL DE BELLAS ARTES, BUENOS AIRES, ARGENTINA

The stately neo-classical building housing the Museo Nacional de Bellas Artes (MNBA) is almost as impressive as the art within, including pieces by most of Argentina's top painters and sculptors. *p227*

MUSEO DEL ORO, BOGOTÁ, COLOMBIA

Discover what drew the conquistadors to South America at this engrossing museum, dedicated to the metalworking skill of Colombia's diverse pre-Hispanic cultures. Room after room is filled with golden wonders; come on Sundays and you can browse this priceless treasure for free. *p229*

SOUTH AMERICA

RUINS THAT WON'T WRECK YOUR BUDGET

Reaching Machu Picchu is a big-budget proposition; try these less costly relics from South America's pre-Columbian civilisations.

SOUTH AMERICA

CHAVÍN DE HUÁNTAR, PERU

The most intriguing of the ceremonial centres in the central Andes, phenomenal Chavín de Huántar is a maze of temple-like structures above ground, and labyrinthine underground passageways below. A palatable entry fee and easy transport push this to the top of the ruins list. *Chavín de Huántar, Peru; entry S15.*

CUEVA DE LAS MANOS, ARGENTINA

Cueva de las Manos' Unesco-listed rock art is remarkable. Recesses in near-vertical cavern walls are adorned with hand imprints, drawings of llama-like guanacos and, from a later period, abstract designs of mysterious significance. *cuevadelasmanos.org; Perito Moreno, Argentina; entry AR$200.*

KUÉLAP, PERU

A manageable entry fee (and easy transport by minibus from Chachapoyas) will get you access to the oval fortress of Kuélap, constructed by the Chachapoyan people between CE 900 and 1100, and surrounded by a stone wall. Save on the cable-car fee by hiking up the 9km path from Tingo Viejo. *Neuvo Tingo, Peru; entry S20.*

RUINS OF TASTIL, ARGENTINA

Reached from the onetime frontier town of Salta, the ruins of Tastil mark one of the most important centres for the pre-Inca Atacameño people. There's no charge to enter the site, a maze of broken walls guarded by cacti, accessible by local bus to the town of Santa Rosa de Tastil. *Free.*

CARAL-SUPE, PERU

Caral-Supe arose in the Supe Valley some 4500 to 5000 years ago, making it one of the world's earliest cities, alongside those in Mesopotamia, Egypt, India and China. A small entry fee gives access to stone pyramids, amphitheatres, ceremonial rooms, altars, adobe complexes and several sunken circular plazas. *Barranca, Peru; entry S11.*

TIERRADENTRO, COLOMBIA

A small entry fee delivers big when it comes to atmosphere in this complex of decorated tombs, well off the beaten track and surrounded by gorgeous mountain scenery. The tombs were hollowed out from CE 500 to CE 900 and decorated with salamanders and geometric murals. *San Andrés de Pisimbalá, Colombia; entry COP$25,000.*

QUILMES, ARGENTINA

Dating from about CE 1000, the fortified city of Quilmes was a complex indigenous town of some 5000 people, whose inhabitants survived contact with the Inca, but not the Spanish. For a modest entry fee, you can explore the remains of stone-walled houses and climb the hillside to a ruined watchtower. *Calchaquí Valley, Argentina; entry AR$70.*

NAZCA WELLS, PERU

The Nazca Lines are best viewed on an expensive plane or balloon trip, but there's no charge to view another great feat of Nazca engineering (see left). The curious, concentric wells of the Acueductos de Cantalloc, near Nazca town, were part of a once-vast irrigation system, supplying the adobe town of Paredones. *Nazca, Peru; free.*

BUENOS AIRES, ARGENTINA

From football to dance and painting to politics, Buenos Aires is a city where emotions run high, finding their best expression in Buenos Aires' lively arts scene. Feel the vibe at free concerts and exhibitions, in public tango halls and out on the streets.

⚓ Caminito

For a dose of Buenos Aires nostalgia, wander down this street of brightly coloured houses – a typical neighbourhood for the Italian immigrant shipyard workers who helped build the capital's fortunes. As you stroll, look out for street performers dancing tango and La Boca's famous Maradona mural. *La Boca; free.*

⚓ Museo Benito Quinquela Martín

Argentinian artists get their moment in the sun at the Quinquela Martín museum, housed in the former home of Benito Quinquela Martín, the most celebrated painter of the port city and the vibrant streets of La Boca. *buenosaires.gob. ar/museoquinquelamartin; Ave Pedro de Mendoza; closed Mon; suggested donation AR$100.*

✋ Museo Casa Carlos Gardel

Even if you don't know much about Carlos Gardel, spend any time in Buenos Aires and you'll soon recognise his face – it's everywhere. The celebrated tango crooner's former home is now an interesting museum, and there are murals of Gardel painted in nearby Calle Zelaya and Calle Agüero. *buenosaires.gob.ar/museos/museo-casa-carlos-gardel; C/Jean Jaurés 735; closed Tue; entry AR$50, free on Wed.*

⚓ Centro Cultural Kirchner

It was former president Néstor Kirchner who, in 2005, first proposed turning the abandoned former central post office into a cultural centre. He died in 2010 before the project was completed, but this breathtaking Beaux-Arts structure now hosts multiple art galleries and events spaces that host regular free events. *cck. gob.ar; C/Sarmiento 151; many events free.*

⚓ Museo de Arte Moderno de Buenos Aires

Get a free midweek dose of culture at the Museum of Modern Art, housed in a former cigarette factory that has been transformed

Below: local artist Benito Quinquela Martín used the buildings of Caminito ('little path') in La Boca as his canvas. Right: Recoleta Cemetery.

into an airy gallery. The walls are decked out with works by Latin American and international artists, from Pablo Curatella Manes, Raquel Forner and Marcelo Pombo to Josef Albers and Wassily Kandinsky. *www. museomoderno.org; Ave San Juan 350; closed Tue; AR$50, free on Wed.*

🖐 Museo Nacional de Bellas Artes
The grand, neoclassical facade of the Museo Nacional de Bellas Artes (MNBA) hides the city's most important fine arts collection, containing key works by Benito Quinquela Martín, Xul Solar, Eduardo Sívori and other Argentine artists, plus work from European masters such as Cézanne, Picasso, Rembrandt and Van Gogh. *bellasartes.gob.ar; Ave del Libertador 1473; closed Mon; free.*

🖐 Recoleta Cemetery
The mausoleums at Recoleta are world famous, but while that of Eva Perón draws the crowds, the cemetery's most elaborate tombs are elsewhere. Look for the mausoleum of newspaper magnate José C Paz (the Argentinian William Randolph Hearst) and the 'tomb' of Dorrego Ortíz Basualdo, with its own chapel. *cementeriorecoleta.com.ar; C/Junín 1760; free.*

🖐 Centro Cultural Recoleta
Part of the original Franciscan convent and alongside its namesake church and cemetery, this brightly muraled cultural centre houses art galleries, exhibition halls and a cinema. It's a great place to check the local cultural pulse; exhibitions are usually free, while tickets to films and shows are reasonably priced. *centroculturalrecoleta. org; closed Mon; C/Junín 1930; free.*

🖐 Casa Rosada
Take a free guided tour of the lavish, salmon-pink President's building and see the balcony from which Juan and Evita Perón once addressed enraptured crowds. Every Thursday at 3.30pm, the Mothers of the Plaza de Mayo, whose children were 'disappeared' during the military dictatorship of 1976–83, gather in front of the Casa Rosada in a moving act of remembrance. *visitas.casarosada.gob. ar; C/Balcarce 50; closed Mon & Tue; free.*

🖐 Usina del Arte
This former power station was turned into a spectacular concert venue to help regenerate a formerly sketchy area of La Boca. It's a gorgeous building and the two concert halls have top-notch acoustics. Best of all, nearly all the art exhibitions, concerts and dance performances here are free. *buenosaires.gob.ar/usinadelarte; C/Agustín Caffarena 1; most events free.*

🖐 Tour BA with a local
Buenos Aires is easy to explore under your own steam, but the city's history comes alive when you learn the stories behind the buildings and monuments. 'Pay what you want' tours starting near Casa Rosada are led by well-informed locals; tip your guide with an amount you think is appropriate. *buenosaireslocaltours.com; tips appreciated.*

FREE TANGO

Don't waste your money on an expensive, tacky tourist dance show; the best place to see real tango is at a *milonga* (dance hall). Many have a small cover charge but there's free entry to one of the most atmospheric, Milonga La Glorieta (11 de Septiembre and Echeverría, Belgrano; Sat & Sun). It's hard to imagine a more romantic setting than this park bandstand where, every weekend, dancers of all ages and levels come to tango together; the sessions usually start with some free tuition.

RIO DE JANEIRO, BRAZIL

It's hard not to fall for Rio de Janeiro, with its gorgeous beaches, rainforest-covered mountains and samba-fuelled nightlife. Resist the urge to splurge on beachfront hotels and high-end restaurants, and hit the cultural centres and monuments for free.

⛎ Centro Cultural Banco do Brasil

Housed in a restored 1906 building, the Centro Cultural Banco do Brasil hosts some of Rio's best exhibitions – all of which are free. Take your time roaming the galleries, or drop by after dark for evening concerts and film screenings. *bb.com.br; Rua Primeiro de Março 66, Centro; closed Tue; free.*

⛎ Escadaria Selarón

One of Rio's best-loved attractions is the tiled staircase connecting Lapa with Santa Teresa. Created by the Chilean-born artist Jorge Selarón, these wildly decorated stairs have become a symbol of Lapa's creative spirit. *Off Rua Joaquim Silva, Lapa; free.*

⛎ Mosteiro de São Bento

Built between 1617 and 1641 on the hilltop of Morro de São Bento, the monastery has a beautiful baroque interior. Come for mass when the monks sing Gregorian chants. *mosteirodesaobentorio.org.br; Rua Dom Gerardo 68, Centro; free.*

⛎ Paço Imperial

This former palace (1743) once housed the exiled royal family of Portugal; now it's been converted into galleries hosting free exhibitions. In the adjoining plaza, Princesa Isabel announced the liberation of slaves in 1888. *amigosdopacoimperial.org.br; Praça Quinze de Novembro 48; closed Mon; free.*

⛎ Bip Bip

A top spot to catch a *roda de samba* (informal samba played around a table). Although the place is just a storefront and some tables, it hosts serious jam sessions as the evening progresses; there's no cover charge, but do tip the musicians. *facebook.com/barbipbip; Rua Almirante Gonçalves 50, Copacabana; closed Sat; free.*

⛎ Instituto Moreira Salles

This cultural centre hosts exhibitions showcasing the works of Brazil's best photographers and artists. The gardens, complete with artificial lake, were designed by Brazilian landscape architect Roberto Burle Marx. *ims.com.br; Rua Marquês de São Vicente 476, Gávea; closed Mon; free.*

Below: there are 215 steps on the Escadaria Selarón, adorned with more than 2000 tiles from 60 countries.

© Pintai Suchaisri | Getty Images

BOGOTÁ, COLOMBIA

Bogotá is Colombia's pulsing heart. During the day, history and culture call, from the atmospheric colonial quarter of La Candelaria to nearly 80 museums (many free) scattered about the city. At night, brace yourself: Bogotá rocks!

✋ Museo del Oro

Bogotá's most famous museum contains more than 55,000 pieces of gold and other objects from all of Colombia's major pre-Hispanic cultures. Most days there's a charge to view these shiny treasures, but Sundays are free to all. *banrepcultural. org/museo-del-oro; Carrera 6, No 15-88; closed Mon; entry COP$4000, Sun free.*

✋ Museo Botero

At the Banco de la República, this massive museum celebrates all things chubby: hands, oranges, women, birds, revolutionary leaders – all, of course, the robust paintings and sculptures of famed artist Fernando Botero. *banrepcultural.org/museo-botero; C/11, No 4-41; closed Tue; free.*

✋ Cine Tonalá

One of Bogotá's best-loved cinemas champions Latin American/Colombian films; the adjoining cultural centre (in a renovated 1930s La Merced neighbourhood mansion) offers a hip bar, cheap Mexican food and rousing club nights. *cinetonala.co; Carrera 6A, No 35-37; films from COP$7000.*

✋ Noche de Galerías

Get drunk on culture? And free wine? Where do I sign up? On four occasions every year, some of Bogotá's coolest galleries open their doors for Noche de Galerías (Gallery Nights), a wine-driven gallery crawl, with free drinks and tours. *nochedegalerias.co; various galleries; free.*

✋ Bogotá Graffiti Tour

This walking tour focuses on Bogotá's urban art scene. The tour itself is free, but a COP$20,000 tip is recommended for the guide. *bogotagraffiti.com; Parque de los Periodistas; free.*

✋ Free music for all

Bogotá is renowned for free concerts such as the Rock al Parque (rockalparque.gov.co; Parque Simón Bolívar; Jun/Jul; free). Music also fills the streets for the 10-day Festival de Verano (idrd.gov.co/festival-verano; various venues; Aug; free).

Below: take a street-art tour of Bogotá to check out the city's ever-changing self-expression.

SOUTH AMERICA

LIMA, PERU

Lima is no mere stopover. Its leafy suburbs and international vibe have made countless travellers hang up their boots. Come for low-cost culture in excellent museums, and some bargain street food.

Museo de Arte de Lima (MALI)
Housed in a Beaux-Arts building, Lima's top fine art museum covers everything from pre-Columbian to contemporary works, offering insights into Chavín and other pre-Incan cultures. On Sundays the admission fee drops to almost nothing. *mali.pe; Paseo Colón 25; closed Wed; entry S30, S1 on Sun.*

Museo Central
Housed in a graceful bank building, the Museo Central provides an overview of several millennia of Peruvian art, from pre-Columbian pottery to the watercolors of Pancho Fierro, which paint a powerful picture of 19th-century Lima society. *bcrp. gob.pe/museocentral.html; cnr calles Lampa & Ucayali; closed Mon; free.*

Museo de la Inquisición
The Spanish Inquisition once carried out its persecutions in the building housing this museum. Morbid wax figures are stretched on racks and flogged; the old first-floor library retains a baroque wooden ceiling that is scarily beautiful. *congreso.gob.pe/ participacion/museo; Plaza Bolivar; free.*

Monasterio de San Francisco
The moderate entry fee at this Franciscan monastery and church allows access to the bone-lined catacombs (containing an estimated 70,000 remains) and its remarkable library housing 25,000 antique texts, some of which pre-date the conquest. Admission includes a 30-minute guided tour. *museocatacumbas.com; cnr calles Lampa & Ancash; entry S15.*

Parque del Amor
With its Gaudí-esque curved benches and arches decorated in mosaics spelling out messages of love, this clifftop space in Miraflores feels like a gallery with no walls, and lovers come here to watch the paragliders launch across Lima's beaches. *Malecón Cisneros; 24hr; free.*

Lugar de la Memoria
An ambitious project to preserve the memory of victims of violence during Peru's tumultuous period from 1980 to 2000, this postmodernist museum features exhibits that commemorate victims and events, aiming to help Peruvians heal. *lum.cultura. pe; Bajada San Martin 151; closed Mon; free.*

Take a free walking tour
Get under the skin of Lima with engaging tours that wind through the historical centre, exploring food stops or roaming around Barranco by night. Tours are free but most people leave a tip. Operators include Strawberry Tours (strawberrytours. com/lima) and Free Walking Tours Peru (freewalkingtoursperu.com).

SOUTH AMERICA

SÃO PAULO, BRAZIL

Enormous and sometimes overwhelming, São Paulo is a vital hub for Brazilian culture, creativity and cuisine. But you don't need big bucks to sample the first-rate museums, cultural centres, experimental theatres and vigorous nightlife.

✋ Museu Afro Brasil

This Parque Ibirapuera museum chronicles 500 years of African immigration (and the 10 million African lives lost in the construction of Brazil) as well as a rotating array of exhibitions. *museuafrobrasil.org.br; Ave Pedro Álvares Cabral, Parque Ibirapuera, Gate 10; closed Mon; entry R$6, free Sat.*

✋ Sala São Paulo

The city's world-renowned concert hall is located in the gorgeously restored Júlio Prestes train station. On Sunday mornings, concerts are free. Grab tickets at the box office on the Monday prior. *salasaopaulo.art. br; Praça Júlio Prestes; times vary; free Sun.*

✋ Museu Xingu

Casa Amarela houses artefacts amassed by Irmãos Villas-Bôas, a three-brother Brazilian activist team who were the first white men to come into contact with the indigenous communities of the Amazon. *casaamarela. art.br/espacos/museu-xingu; Rua José Maria Lisboa 838, Jardins; closed Sat & Sun; free.*

✋ Edifício Copan

Designed by Oscar Niemeyer, this building, with its serpentine facade and *brises soleil* (permanent sunshades), is the city's most symbolic piece of architecture. Its spectacular rooftop opens only on weekdays, for a 15-minute visit. *copansp.*

com.br; Ave Ipiranga 200, República; closed Sat & Sun; free.*

✋ Mosteiro de São Bento

Among the city's oldest and most important churches is São Bento, dating to 1598. Step inside to view the riotously decorated interior and stained glass. *mosteiro.org.br; Largo de São Bento, São Bento; free.*

✋ São Paulo Free Walking Tour

Condensing more than 450 years of history into a couple of hours, these weekend walks are a great way to get to know the old centre of São Paulo, and the Avenida Paulista and Vila Madalena neighbourhoods. *spfreewalkingtour.com; Praça da República; Sat & Sun; free, but tips appreciated.*

Below: the interior paintings of the Mosteiro de São Bento were inspired by Byzantine art.

FOOD & DRINK

South American food is getting noticed around the world – particularly the meaty cuisines of Brazil and Argentina – but for most travellers, a trip to the continent is a voyage of culinary discovery. You can pay a lot for a prime steak and a bottle of Mendoza red in top restaurants, or find inexpensive *parillas* (steakhouses) serving prime cuts at burger prices.

By eating strategically – a filling breakfast, low-cost set-menu lunches and evening street-food snacking – you can get by on less than US$20 per day and still have money left for occasional evening splurges.

Above and right: buying ingredients from local markets rather than eating out every night is a great way to save some money.

SELF-CATERING & STREET FOOD

Self-catering in South America is easier than you might think, thanks to the abundance of hostels with kitchens. Indeed, self-catering in the evening can halve your dining costs, as evening meals are almost always more expensive than lunches. Shop at open-air markets to further shrink your ingredients bill.

South America's other great food bargains are out on the street, in stalls clustered around public plazas, markets and transport hubs. Some treats may already be familiar – *empanadas* (meat pasties), Brazilian *pastéis* (deep fried crispy pies), *choripán* (chorizo sandwiches) – while snacks such as *anticuchos* (grilled heart kebabs) are waiting to be discovered; but all are cheap, filling and satisfying.

• **Follow the lead of locals** – good stalls are busy, bad stalls are empty; ask your hotel, taxi driver or shop staff for recommendations.

• **Food on the move** – for long-distance bus and train rides, grab some

empanadas (or similar portable snacks) to go.

• **Time your treats** – street stalls serve throughout the day, with markets and business districts being busiest at lunchtime, and neighbourhood eat-streets and drinking hubs getting busy as dusk falls.

SIT-DOWN DINING

If you know the cheap seats, sit-down dining doesn't have to mean expensive dining. Make lunchtime the big meal of the day; many restaurants offer an inexpensive *menú del día* – usually something filling that they know they'll sell lots of, often with juice and *sopa* (soup) on the side. You'll also hear the terms *ejecutivo* and *almuerzo*. Get used to lots of stews, rice and beans and you'll thrive.

• **Frugality loves company** – cut evening restaurant costs by dining with a friend; plates for two are normally cheaper than two mains.

• **Make meat count** – prime cuts are costly; stick to stews or smaller portions beefed up with starch in the form of rice and *arepas* (corn cakes).

• **Street food chasers** – instead of big, multi-course dinners, grab something light (a salad or soup) and top up with cheap street food at the most interesting time of day to be out on the street.

SMART DRINKING

It would be remiss to leave South America without sinking a pisco sour in Peru, a caipirinha in Brazil, or a glass of Malbec in Argentina, but costs can mount. Get a takeaway from a supermarket to stretch your centavos. Don't overlook inexpensive and tasty non-alcoholic drinks – juices, *mate* (tea made from yerba mate leaves), coca tea (yes, made from that coca) and the continent's famous coffee.

HIGHLIGHTS

PALACIO DEL VINO, SANTIAGO, CHILE

Get the Chilean wine buzz without the financial hangover in this popular 'wine palace', set in a historic building in Barrio Brasil. Mix tasty gran reserva wine with an afforda-ble, seafood-rich menu and you have budget fine dining on demand. p236

MERCADO CENTRAL, QUITO, ECUADOR

Quito's fastest and feastiest food is served out on the street at the city's central market, where you can graze for pennies on *locro de papas* (potato, cheese and avocado soup), *empanadas* and *fritada* (fried pork with hominy). Bring an appetite not a big budget. p238

SOUTH AMERICA

BUENOS AIRES, ARGENTINA

Not content with being a bastion of culture, the Argentine capital also scores highly for thrifty eating, with cost-effective steaks, bargain choripán (sausage sandwiches) and inexpensive Italian food, thanks to the ítalo-argentinos who flocked here in the 1850s.

SOUTH AMERICA

✋ Sip morning mate at Plaza Lavalle

Make like a local and pick yourself up some *mate* (loose-leaf tea), a drinking gourd and *bombilla* (metal straw) then head over to Plaza Lavalle with a vacuum flask of hot (but not boiling) water, and watch the city come awake while you sip your morning brew. *Free.*

✋ Plaza Francia Feria Artesanal

The weekend artisan market in Plaza Francia does a brisk trade in handicrafts, but keep an eye out for sellers of homemade *pan relleno* (bread stuffed with cheese and tomato), one of the cheapest and most filling meals you'll find in town. Look for vendors wandering among the stalls carrying baskets of their tasty wares. *Recoleta; Sat & Sun; pan relleno AR$20-25.*

✋ Pizzería Güerrin

Forget artisan pizzas; the Argentinian version is a fat, doughy, greasy, satisfying carb-fest. Eat slices standing up at the counter for a snip of the price of sitting at a table, and ramp up the calories with a chaser of *fainá*, a chickpea-based flatbread. *guerrin.com.ar; Ave Corrientes 1368; pizza from AR$25 per slice.*

✋ El Sanjuanino

This long-running, cosy little joint has some of the cheapest food in Recoleta. Order spicy *empanadas*, *tamales* or *locro* (a spicy stew of maize, beans, beef, pork and sausage). Many take their food to go – Recoleta's lovely parks are just a couple of blocks away. *facebook.com/ elsanjuaninoempanadas; C/Posadas 1515; mains AR$110-220.*

✋ El Banco Rojo

A San Telmo youth magnet, this trendy joint serves up sandwiches, falafels, burgers and tacos, as well as moderately priced beers and spirits, and a friendly, grungy vibe. Check the chalkboard menus and try the *empanada de cordero* (lamb pasty) if it's on; it's a feast. *bancorojo.com. ar; C/Bolívar 866; mains AR$115-150.*

Below: classic Argentinian street food, a traditional *choripán* sandwich filled with chorizo sausage, tomato and chimichurri sauce. **Right:** a seafood ceviche.

© Alexandr Vorobev | Shutterstock

Chori

Elevating the humble Argentine *chori* (sausage) to new heights, this hip joint has bright yellow walls decorated with smiling cartoon sausages. The quality, 100% pork *choris* and *morcillas* (blood sausages) hang on display – choose from a range of gourmet toppings and homemade breads for a fast *choripán* sandwich. *facebook. com/Xchorix; C/Thames 1653; choripán from AR$150.*

Nola

The chef at this casual and wildly popular Louisiana-style gastropub was New Orleans born and raised, meaning you can take a break from Argentinian food with some fine Cajun cuisine, from fried chicken sandwiches to spicy gumbo. Craft beer, cocktails and wines by the glass, too. *nolabuenosaires.com; C/Gorriti 4389; mains AR$140-250*

Parrilla Peña

This simple, traditional, long-running *parrilla* (steakhouse) is well-known for its quality cuts and generous portions. Meat here is cooked to perfection, and there's a good wine list. Check out the inventive system the staff use to retrieve bottles from the highest shelves. *C/ Rodríguez Peña 682; closed Sun; steaks from AR$300.*

Gran Dabbang

The rule-breaking, experimental fusion food conjured up by Mariano Ramón can be sampled in the small, packed dining room of this unassuming restaurant. Order small plates to share and get ready for a delicious mix of flavours drawn from Ramón's travels. *facebook. com/grandabbang; Ave Scalabrini Ortiz 1543; closed Sun; small plates from AR$120.*

Chan Chan

Thanks to fair prices and quick service, this colourful Peruvian place is usually packed with hungry souls devouring plates of ceviche (seafood cured in citrus) and *ajiaco de conejo* (rabbit and potato stew). Come early or prepare to wait for a table at night. *facebook.com/ chanchanbsas; Ave Hipólito Yrigoyen 1390; mains from AR$170.*

Cadore

Cadore, one of BA's classic *heladerías* (ice-cream parlours), was founded by the Italian Olivotti family in 1957. Famous for its version of dulce de leche (milk caramel) ice cream, the place gets busy with theatre crowds late into the night. *heladeriacadore.com.ar; Ave Corrientes 1695; ice cream from AR$120.*

El Cuartito

The classic Buenos Aires fusion – a cold beer, a fat slice of pizza and an *empanada*. Enjoy standing up at the counter in one of Buenos Aires' oldest pizzerias, alongside local office workers and longtime neighbourhood residents, with a good view of the old sports posters and football shirts hanging on the walls. *C/Talcahuano 937; closed Mon; pizza from AR$150.*

THE TASTE OF HOME

When *porteños* (Buenos Aires locals, literally 'port-dwellers') return home from their travels, one of the first places they'll visit is the Costanera Sur for *choripán* (chorizo sandwich) or *bondiola* (braised pork shoulder) for AR$25 to AR$30. This is street food Buenos Aires-style: a long line of roadside parrillas on Ave Int Hernán M. Giralt sell filling, meaty sandwiches for bargain prices. The smell of barbecuing meat – perhaps the essence of the city – is difficult to resist.

SOUTH AMERICA

SANTIAGO, CHILE

Surprising, cosmopolitan, energetic, sophisticated and worldly wise, Santiago is a city of madhouse parties and top-flight restaurants that charge economy prices. Devote the days to free museums and save your money for meals to remember

🖐 La Diana

Built within the walls of an old monastery, La Diana defies easy description. Its ceilings are adorned with as many potted plants as chandeliers, its tables are a mishmash of found furnishings, and its menu ranges from grilled Chilean fish to seafood-packed pizzas. *ladiana.cl; C/ Arturo Prat 435, Centro; closed Mon; mains CH$6000-8500.*

🖐 Salvador Cocina y Café

This lunch spot packs a surprising punch, with market-focused menus that highlight unsung dishes from the Chilean countryside. Chef Rolando Ortega won Chile's chef of the year award in 2015 and the tables have been full ever since. *facebook.com/SalvadorCocinaYCafe; C/ Bombero Ossa 1059, Centro; closed Sat & Sun; mains CH$7700.*

🖐 Fuente Suiza

Seriously good *lomo* (pork) sandwiches and flaky deep-fried empanadas make this simple family-run restaurant the perfect place to prepare for (or recover from) a long night of drinking. Meals are starchy, meaty and satisfying. *fuentesuiza.cl; Ave Irarrázaval 3361, Ñuñoa; sandwiches CH$3200-6300.*

🖐 Palacio del Vino

This down-to-earth 'wine palace' gets high marks for both atmosphere and price. Not only is it located in a character-rich heritage building in Barrio Brasil, but you can get a glass of gran reserva wine for CH$3000 and a seafood dish to go with it for less than CH$6000. *palaciodelvino. cl; Ave Brasil 75, Barrio Brasil, closed Sun; mains CH$5000-8000.*

🖐 Café Bistro de la Barra

Worn old floor tiles, a velvet sofa, a 1940s swing and light fittings made from cups and teapots make a quirky-but-pretty backdrop for some of the best lunches and *onces* (afternoon teas) in town. The rich sandwiches include salmon-filled croissants or Parma ham and rocket on flaky green-olive bread, but save room for berry-drenched cheesecake. *cafedelabarra.cl; JM de la Barra 455, Barrio Bellas Artes; sandwiches CH$4000-5500.*

🖐 Plaza Garibaldi

From the saloon-style doors to the tacos, quesadillas and *chimichangas* (fried burritos), Plaza Garibaldi will transport you a few countries north to Mexico. Spice levels are dialled down slightly for Chilean palates, but food is fresh and filling. *facebook.com/ pgaribaldi; C/Moneda 2319, Barrio Brasil; closed Sun; mains CH$5000-9000.*

MEDELLÍN & THE ZONA CAFETERA, COLOMBIA

Colombia's second city is a much-loved retreat for digital nomads, expats and travellers. Credit its cool climate, international outlook, inexpensive food and easy access to the highlands, including the Zona Cafetera plantations that produce Colombia's coffee.

Restaurante Itaca, Medellín

This hole-in-the-wall restaurant on the edge of downtown prepares fantastic, fairly priced Colombian plates. At lunch there are a couple of set meals. *facebook. com/RestauranteItaca; Carrera 42, No 54-60; set lunch COP$14,000, mains from COP$16,000.*

Arte Dolce, Medellín

Lorenzo the confectioner evokes true Italian gelato; our favourite flavour is the Mediterranean, which blends pistachios, almonds, lemon, orange and olive oil. *facebook.com/artedolcemde; Carrera 33 No 7-167; gelato COP$4500-8000.*

Betty's Bowls, Medellín

Come to Betty's Bowls for beautifully presented granola bowls with different fruit combos, plus fresh juices and toast with savoury toppings. *bettysbowls.com; Carrera 32D, No 7a-77; mains from COP$10,000.*

Helena Adentro, Filandia

Happy, hip restaurant serving innovative takes on traditional Colombian cuisine, made with fresh ingredients from local farms. The menu is constantly changing but it's all delicious and very reasonably priced considering the quality. *helenaadentro.com; Carrera 7, No 8-01, Filandia; mains from COP$23,000.*

El Jardín de las Delicias, Manizales

Modern Colombian dishes with an international twist crafted by chef Pablo, who blends cooking techniques learned in Paris with local ingredients. Enjoy your meal on the sunny terrace. *facebook.com/ eljardindelasdelicias.rest; Carrera 25, No 68-19, Manizales; closed Sun; mains from COP$27,000.*

El Rincón de Luci, Salento

We love Luci for her great-value Colombian meals: hearty fish, beef or chicken dishes served with rice, beans, plantain and soup, consumed at communal tables next to strangers who may become friends. *Carrera 6, No 4-02, Salento; breakfast COP$7000, lunch & dinner COP$9000.*

Below: a Colombia classic, *trucha al ajillo*: river trout served with fried plantain and garlic sauce.

SOUTH AMERICA

QUITO, ECUADOR

High in the Andes, Quito is where traditional Ecuadorian culture melds into vibrant and sophisticated big-city living. Target the free sights and save your money for empanadas, ceviche and other food treats from the streets.

Mercado Central

For stall after stall of some of Quito's most traditional (and cheapest) foods, head straight to the central market and try everything from *locro de papas* (potato soup with avocado and cheese) to *yaguarlocro* (potato and blood-sausage soup) and *fritada* (fried chunks of pork, served with hominy). *Ave Pichincha, Old Town; meals E$1.50–4.*

Cafetería Modelo

Opened in 1950, Modelo has long been a great spot to try traditional snacks such as *empanadas de verde* (made with plantain dough), *quimbolitos* (cake-like corn dumplings) and tamales. The fun, slightly kitsch decor makes the wait for food interesting. *facebook.com/ cafeteriamodeloec; C/Sucre 391, Old Town; mains E$4-6.*

Masaya Bistro

This large, inviting space with long wooden tables spills onto the garden of the Masaya Hostel. Ecuadorian-international food is served up in casual, relaxed surroundings: beef in chimichurri sauce, quinoa croquettes, Ecuadorian-style ceviche, burgers and more. *masaya-experience.com/quito; cnr calles Venezuela & Rocafuerte, Old Town; mains E$3-8.*

Cosa Nostra

Italian-owned Cosa Nostra has nearly three-dozen varieties of pizza stacked high with generous toppings – many consider these the best pizzas in town. There's also good gnocchi, pasta, and (of course) tiramisu for dessert. *pizzeriacosanostra.ec; C/República del Salvador 34-234 y Moscú; mains E$11-18.*

Fried Bananas

For innovative Ecuadorian food at a reasonable price, check out this atmospheric restaurant where the namesake bananas flavour dishes such as delectable shrimps with fried bananas and vodka. *facebook.com/ friedbananasrestaurante; C/Foch E4-150, New Town; closed Sun; mains E$6-10.*

Altar Cervecería

A hipster haven of polished concrete and blonde wood, this new bar has some tasty beers from its own brewery as well as a selection of other Ecuadorian and regional craft beer on tap. The food – burgers, chicken wings, beer-battered fries (E$5.50 to E$8) – is excellent and there's regular live music. *cnr calles Olmedo & León; closed Sun & Mon.*

SOUTH AMERICA

LIMA, PERU

Street snacks in Lima – papa rellena (fried potato stuffed with mince), picarones (deep-fried sweet potato and squash), anticuchos (beef heart barbecued on skewers) and more – are all dirt cheap, and it doesn't cost much more for a sit-down meal.

🖐 ámaZ

Raise your dinner budget for Chef Pedro Miguel's Amazon-tribute cuisine. Start with tart jungle-fruit cocktails and *tostones* (plantain chips) then enjoy lip-smacking banana-leaf wraps, aka *juanes*. *amaz.com. pe; Ave La Paz 1079; mains from S20.*

🖐 El Bodegón

It's value rather than low prices that lures people to this taverna serving Peruvian home-style cooking. Share several dishes; we rate the delectable *rocoto relleno* (stuffed pepper). *elbodegon.com.pe; Ave Tarapaca 197; mains from S28.*

🖐 Barra Chalaca

If you need a tonic for the city crush, head to this diner-style café and order *curatodo* (cures everything) – a fishbowl of tropical juices and fresh herbs. Then follow up with *tiradito chucuito* (sashimi with capers, avocado and garlic). *barrachalaca.pe; Ave Camino Real 1239; mains from S14.*

🖐 La Panetteria

This neighbourhood gem serves a variety of breads, from baguettes to experiments such as pesto or *aji* (Peru's slightly spicy pepper) loaves and gorgeous pastries. It's busy on weekends. *facebook.com/ lapanetteriabarranco; Ave Grau 369; closed Mon; baked goods from S8.*

🖐 Café de Lima

Sitting on the corner of a busy Miraflores avenue, this cafe is a contemporary ode to traditional Lima. Come for baskets of freshly baked goodies and classic Lima desserts such as *suspiro de limón* (lemon cream, here with a hint of lemongrass). *cafedelima. pe; Ave Angamos 1003; baked goods from S6, mains from S25.*

🖐 Choco Museo Miraflores

More a shop than a museum, with free tours presenting the history of chocolate. Explore the ingredients before diving into the taste-testing, with free samples of raw cacao, sublime chocolate bars, chocolate tea and fair-trade hot cocoa. *chocomuseo.com; C/ Berlin 375; admission free.*

Below: tasty street-food snacks in food-loving Lima.

SOUTH AMERICA

WINE TASTING ON A BUDGET IN SOUTH AMERICA

South American wine regions are romantic destinations, but you don't have to kiss goodbye to all your pesos to sample the fruit of the harvest.

FINCA NARBONA – URUGUAY

The wineries around Carmelo are the places to sample little-known Tannat. You don't need to be an overnight guest at Narbona Wine Lodge to see the impressive winery: just come for lunch with wine pairings. *narbona. com.uy; Carmelo, Uruguay; meals form US$25.*

VENEZUELA

COLOMBIA

ECUADOR

GUYANA

SURINAME

FRENCH GUIANA

CAFAYATE WINE REGION – ARGENTINA

Fewer tourists make it to Cafayate, providing whites to rival Mendoza's legendary reds (including the aromatic white varietal Torrontés), but it's an easy side-trip from Salta, with plenty of wineries offering economical tastings. *welcomeargentina. com/cafayate.*

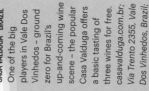

VIÑA CONCHA Y TORO - CHILE

The largest wine producer in Latin America may not be especially quaint, but you can at least get to it on the metro from downtown Santiago. *conchay toro.com; Ave Virginia Suber caseaux 210, Pirque, Chile; tastings CH$18,000.*

CASABLANCA VALLEY - CHILE

You came to Valparaiso for picturesque architecture and Pablo Neruda, now take a day tour with Wine Tours Valparaiso – the perfect way to see the nearby Casablanca Valley. winetoursvalparaiso. cl: Cerro Alegre, Valparaiso, Chile; tours from CH$50,000.

MONTGRAS WINERY - SANTA CRUZ, CHILE

Make a day of it at this beautiful winery located in Chilean wine region Colchagua Valley; in addition to tasting Carménère, you can go hiking or mountain biking. montgras. cl: Camino Isla de Yáquil s/n, Palmilla; tasting with food pairing CH$15,000.

MENDOZA ON TWO WHEELS - ARGENTINA

Rent a bicycle and explore *los caminos del vino* (roads of wine) close to the city of Mendoza. Mr. Hugo Bikes offers bicycles, lockers and maps, and organises tours. *facebook. com/mrhugobikes; Urquiza 2288, Maipú, Mendoza; prices vary.*

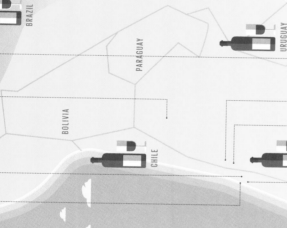

SALENTEIN WINERY - VALLE DE UCO, ARGENTINA

In Argentina's buzzed-about Uco Valley, get more bang for your peso by timing winery visits with the cultural events on schedule at Salentein's amphitheatre. *bodega salentein.com; RP 89, Los Árboles, Tunuyán, Argentina; closed Sun; tastings AR$250.*

CASA VALDUGA - BRAZIL

One of the big players in Vale Dos Vinhedos – ground zero for Brazil's up-and-coming wine scene – the popular Casa Valduga offers a basic tasting of three wines for free. *casavalduga.com.br; Via Trento 2355, Vale Dos Vinhedos, Brazil; free...*

PERU

BOLIVIA

BRAZIL

PARAGUAY

CHILE

URUGUAY

ARGENTINA

FESTIVALS & EVENTS

Not to be outdone by its northern counterpart, South America does festivals on an epic scale, with Carnaval in Rio de Janeiro making other carnivals look like village fetes. But festival time isn't always as easy on the pocket as it is on the ear.

It is possible to get through Rio's carnaval without blowing your bank balance, you won't be so lucky when it comes to flights and accommodation. Savvy travellers fit in festivals as part of a longer trip to avoid peak-price flying dates.

Above: Semana Santa, here in Peru, is a big deal in South America. Right: Carnival celebrations in Rio de Janeiro and throughout Brazil are equally huge.

PRACTICALITIES

The South American festival calendar leans heavily towards Christian feast days, meaning a huge turnout (and peak demand for transport and accommodation) during every big religious celebration, from Christmas to Holy Week. Getting anywhere during Easter can be a slow, expensive ordeal anywhere in the continent – renting a car will liberate you from the scrum for bus and plane seats.

• **Secure your seat** – many flights run only once a day, or a few times a week, so book seats up to six months in advance for major celebrations.

• **Listen for free** – tickets for big music and sporting events sell out fast; don't overlook free festivals such as Bogotá's Rock al Parque (rockalparque.gov.co; June/July), where getting tickets isn't an issue.

• **Watch your stuff** – festivals offer rich pickings for pickpockets and bag-snatchers; leave your valuables safely in your hotel before joining the crowds.

RELIGIOUS FESTIVALS

Even festivals that pre-date the arrival of Christianity have been reinvented with an air of Biblical theatre in devout South America. As well as region-wide events such as Semana Santa (Holy Week), look out for smaller local festivals linked to the patron saints of holy sites, towns and churches.

The big perk of religious festivals is the entry fee: you'll rarely need to pay a cent for South America's most spirited celebrations. Many involve an element of pilgrimage, meaning some intense experiences for those willing to walk and camp alongside the crowds.

MUSIC & CULTURE

South America is a major stop on the global music festival circuit, with stages hosting every imaginable genre, from electronic music to punk and death metal.

Commendably, the continent also throws on plenty of free music bashes to counteract expensive, mobbed mega-fests such as Rock in Rio (rockinrio.com; Sep). Try Bogotá's ticket-free Rock al Parque (rockalparque.gov.co; Jul), Jazz al Parque (jazzalparque.gov.co; Sep) and Salsa al Parque (salsaalparque.gov.co; Nov).

FOOD

South America's food festivals are a feast to rival anything in Europe, Asia or beyond. Gastronomes start drooling at the very mention of Peru's Mistura (Lima, Sept) and Brazil's Brasil Sabor (brasilsabor.com.br; May). Entry fees (where they apply) are at snack prices, and food stalls offer abundant cheap eats.

SPORTS

South Americans love sport – and football in particular – with almost religious devotion. You won't find a cheap ticket to international tournaments; you may do better in the minor and youth leagues, where the next generation of Peles and Messis are coming through.

HIGHLIGHTS

CARNAVAL, RIO DE JANEIRO

The festival by which all others are measured, Rio's annual extravaganza is a riot of feathers, sequins, brocade and gyrating human flesh. For two weeks before Ash Wednesday, neighbourhoods fill with street parties, music and dancing fill the air with noise, and everyone goes slightly bonkers. *p244*

SEMANA SANTA, PERU

Peru's most vibrant event sees the streets carpeted with flowers for Holy Week, while parades and processions haul effigies and full-sized wooden crosses through cities, towns and villages. Expect lots of colour and costumes. After this build-up, Easter itself can feel almost anti-climactic. *p245*

FESTIVALS

✋ Brasil Sabor, Brazil

Brazil's biggest palate-pleaser takes place in May, when more than 700 restaurants in 65 cities offer special plates, promo prices and hard-to-find delicacies. The focus changes from year to year, but authentic Brazilian dishes are the centre of attention. *brasilsabor.com.br; May; free.*

✋ Buenos Aires International Festival (FIBA), Argentina

Buenos Aires' biggest celebration of theatre features shows across 40 venues – some free, some ticketed – riffing on contemporary themes such as diversity and the environment. Expect lots of international talent as well as productions by local companies. *buenosaires.gob.ar/fiba; dates vary; free.*

✋ Carnaval, Rio de Janeiro, Brazil

Sure, hotel and flight prices peak, but there are many ways to join in the fun without spending a lot of cash. In the weeks leading up to the big event, *blocos* and *bandas*, aka roving street parties, happen all across town (see p246). There are also free concerts, and parties at the samba schools that host the big parade. *rio-carnival.net; two weeks leading up to Ash Wednesday; Rio de Janeiro, Brazil; free.*

✋ Feira Rio Antigo, Rio de Janeiro, Brazil

If you're in Rio on the first Saturday of the month, don't miss this fair – there's live music, food and drink vendors and countless stalls selling crafts, clothing and antiques. It all goes down on one of Lapa's oldest streets, Rua do Lavradio, which is lined with picturesque buildings. *Rua do Lavradio, Lapa; 1st Sat of month; Rio de Janeiro, Brazil; admission free.*

✋ Fiesta de la Vendimia, Chile

The grape harvest in the Maipo Valley is a big deal for wine lovers. In April, wineries and towns across the region mark the season with wine-queen pageants, grape crushing, special meals and free events, and of course, plenty of previous years' vintages are consumed. *Maipo Valley, Chile; Apr.*

✋ International Festival of American Renaissance and Baroque Music, Bolivia

The Jesuit Mission towns of the Amazon jungle erupt in glorious harmony every two years for this free celebration of religious music, centred on the city of Santa Cruz de la Sierra. Over 1000 musicians gather to play music that was

Below: the Inti Raymi solstice celebrations are the biggest annual event in Cuzco. Right: tango takes over Buenos Aires in August.

written centuries ago in these same religious establishments. *Chiquitania, Bolivia; Apr/May; most events free.*

⚑ Inti Raymi, Peru

This Inca celebration of the winter solstice honours Inti, god of the sun. Expensive tickets grant entry to the main event in the Sacsayhuaman ruins, but you can view from the mountain above for much less, or forgo the main show and just take in the streets of Cuzco filling with indigenous people in traditional costume. *Cuzco, Peru; June; main show tickets from US$100.*

⚑ Mistura, Peru

Peru's biggest feast fills the streets of Lima with food stalls and delicious smells as locals come together to celebrate Peruvian cuisine. Bring an appetite and graze your way through the city, in between watching cooking competitions, demonstrations and more. It's a chance to sample little-known regional delicacies, too. *Lima, Peru; Sep; tickets from S17.*

⚑ Pase del Niño Viajero, Ecuador

The infant Jesus gets special attention in this high-spirited fiesta on Christmas Eve. Parades fill the streets of Cuenca with floats, wise men on Harley Davidsons, parades of statues, children in costume and cars transformed into mobile kitchens with piles of produce and cooked food. *Cuenca, Ecuador; 24 Dec; free.*

⚑ Réveillon, Rio de Janeiro, Brazil

Give frosty Times Square or rainy London a miss and head to the tropics to ring in the New Year, dressed in white. Rio throws a truly incredible bash, with some two million people taking to Copacabana Beach to watch fireworks light up the night sky. *Copacabana Beach, Rio de Janeiro, Brazil; 31 Dec; free.*

⚑ Semana Santa, Peru

People across Peru get slightly mystical for the final build up to Easter, when Peruvian legend dictates that no sins can be committed. Penitents parade, crucifixes are hauled, streets are carpeted with flowers and fireworks pop, particularly in Lima, Cuzco and Ayacucho. *Across Peru; Mar/Apr; free.*

⚑ Tango Buenos Aires Festival y Mundial, Argentina

The world's biggest tango festival fills the Argentine capital with passion and poses every August, when thousands of couples descend on the city and the streets ooze sensuality. There are free performances, classes, concerts and demonstrations, plus the biggest dance competition of the year, the Mundial de Tango. *buenosaires. gob.ar/tangoba; Buenos Aires, Argentina; Aug; free.*

⚑ Virada Cultural, São Paulo, Brazil

This non-stop, city-wide festival of culture and music, usually held in May in and around some of Centro's best-known public squares, includes circus and theatre performances, dance, music of all kinds and more. The best thing? It's all for free. *capital.sp.gov.br/turista/atracoes/eventos/virada-cultural; São Paulo, Brazil; free.*

EASTER THE SOUTH AMERICAN WAY

In devoutly Catholic South America, Semana Santa (Holy Week) is bigger than Christmas, and there's a certain occult flavour to some of the celebrations. In parts of Brazil and Venezuala, people flog and burn straw effigies of Judas as a gesture of outrage at the betrayal of Jesus, while in Cusco, Peru, a statue of Christ as El Señor de los Temblores (Lord of Earthquakes) is garlanded with firecrackers and paraded through the city.

SOUTH AMERICA

HOW TO PARTY AT CARNAVAL IN RIO

You don't need to splash cash to celebrate Carnaval. Some of the best events – especially Rio's festive street parties – are free.

SALGUEIRO SAMBA SCHOOL

One of the most exciting places to be in the build-up to Carnaval is at a samba school rehearsal, and Salgueiro is a favourite. Don your red and white (Salgueiro's colours) and plan a late night of dancing! *salgueiro. com.br; Rua Silva Teles 104, Andaraí; Sat; admission from R$30.*

MANGUEIRA SAMBA SCHOOL

Every Saturday night from September to Carnaval, this traditional samba school throws a great party. Expect heavy percussion, flowing caipirinhas and a festive crowd. *mangueira.com.br; Rua Visconde de Niterói, Mangueira; Sat; admission from R$40.*

BANDA DE IPANEMA

One of the best roving celebrations in Ipanema happens twice during Carnaval. You can don a costume (or not), and join the massive crowds as they dance through the neighbourhood. *ipanema.com.br/ carnival/banda.htm; Praça General Osório, Ipanema; Sat & Tue of Carnaval & Sat two weeks prior; free.*

MONOBLOCO

The festivities in Rio don't end abruptly on Ash Wednesday. You can join half a million revellers at this huge downtown afterparty hosted by street band Monobloco to bid adeus to Carnaval for another year. *monobloco.com.br; location varies; 1st Sun after Carnaval; free.*

BANDA CARMELITAS

There's nothing quite like shimmying your way through the cobblestone streets of Santa Teresa – preferably dressed as a nun (the costume of choice). *Cnr Rua Dias de Barros & Ladeira de Santa Teresa, Santa Teresa; Carnaval Fri & Tue; free.*

CORDÃO DA BOLA PRETA

Arrive early to join Rio's oldest and biggest street party, with some two million joining the celebration. Costumes are encouraged – especially something creative with black-and-white spots. *cordaodabolapreta. com; Primeiro de Março near Rua Rosário, Centro; Carnaval Sat; free.*

PARADE OF CHAMPIONS

If you don't want to pay for the big parades during Carnaval, come the Saturday just after the big week, when the top six samba schools march through the Sambódromo. It's a dazzling performance, well worth catching. *Sambódromo; Sat after Carnaval; tickets from R$120.*

SAMBA LAND

This huge square near the Sambódromo transforms into party central during Carnaval (and the weekend before), with a wide range of live music throughout the night, plus abundant food and drink vendors. *Praça Onze, Centro; Fri-Tue of Carnaval; admission around R$40.*

RIO FOLIA

During Carnaval, head to the plaza in front of the Lapa Arches for some of the city's best free shows. There's great dancing and people-watching at these open-air concerts. *Praça Cardeal Câmara, Lapa; Fri-Tue of Carnaval; free.*

OUTDOORS & ADVENTURE

From the jungles of the Amazon to the mountains of Patagonia, South America is a continent that's made for adventure; indeed, there's so much wilderness out there that even driving from town to town can be an off-road expedition.

The problem from a budget perspective is not entry fees but access. Getting to many of the region's parks and reserves involves long-distance travel on dirt roads and jungle rivers, putting many out of reach for budget travellers.

Above: hiking in Manu National Park, Peru. Right: a boat tour on the Amazon near Iquitos.

NATIONAL PARKS & NATURE RESERVES

More than 300 national parks protect vast swaths of South America – and they're needed, with threats ranging from forest clearance to poaching. But the wildlife-spotting opportunities are still unparalleled – some parks offer almost guaranteed sightings of jaguars, river dolphins, anacondas and more.

However, the experience comes at a price. Almost all state-managed parks charge an entry fee, though prices are rarely ruinously high – it's the physical process of getting to them (by jeep, on organised treks, by boat or by light aircraft) that adds expense. Some parks also require travellers to take a guide – understandable with jaguars about, but it's another cost to bear in mind.

For cheaper spotting, look to state-managed forests (many edge onto big cities) and smaller nature reserves and conservation areas managed by international charities and indigenous communities. Most charge fees, but offer access to wildlife in a smaller area, meaning more sightings for your peso.

• **Put in the time** – in any reserve, you'll need to stay several days to maximise the chances of meeting the wildlife; some parks reduce the daily fee for longer stays.

• **Almost a park?** – in Patagonia and

the coastal deserts of Peru and Chile, the empty areas between the parks are almost as impressive as the parks themselves, and they're free to explore.

• **Free boat safaris** – river journeys to jungle towns double as rainforest safaris; watch the forest for monkeys and scan the water for anacondas, caiman and capybara.

ADVENTURE ACTIVITIES & ORGANISED TRIPS

With so much rock, water and ice bursting out of the landscape, South America is one big jungle gym. Skiers, trekkers and climbers can tackle Patagonia and the Andes (though you may need to trek to reach the slopes); nature buffs have the Amazon; and surfers will find superlative breaks on the coasts of Brazil, Argentina, Chile and Peru (bring your own board and every break is free).

Organised treks to Machu Picchu, boat trips, wildlife safaris, thrill sports and guided tours are easy to find, but they're rarely cheap (US dollar prices are the norm). For budget-conscious adventures, look to buffer zones outside national parks that you can access independently on foot, by bike or by rental car.

• **Camp the route** – many wilderness areas can only be reached on foot: bring your own camping equipment and trek in independently, saving the cost of an organised trek.

• **Drive yourself** – some driving routes, particularly in Patagonia, cut through scenery to rival any national park; just be ready for dirt roads and long gaps between fuel and food stops.

• **Be a desert dweller** – the coastal deserts of Peru and Chile are sparsely inhabited – and free to explore if you have your own transport, or gear to trek and camp unsupported.

HIGHLIGHTS

LAGO TARAPOTO, COLOMBIA

Taking a *peque-peque* (wooden boat) to reach this gorgeous jungle lake fed by the Río Amazonas is a bargain Amazon adventure. You can swim in the same clear waters where pink dolphins and manatees splash, surrounded by an endless carpet of rainforest. p250

HIKING FROM EL CHALTÉN, ARGENTINA

Argentina's top trekking hub, El Chaltén is the gateway to Parque Nacional Los Glaciares (North), but the free hikes that start at the town limits are the match of almost anything you'll find inside the reserve, with the promise of a warm bed and hot meal when you return. p255

SOUTH AMERICA

THE AMAZON

One of the world's biodiversity hotspots, the world's largest river snakes through dense jungles in Peru, Bolivia, Venezuela, Colombia, Ecuador and Brazil. With this much nature on display, even free adventures are out of this and any other world.

Reserva Biosférica del Beni, Bolivia

Created by Conservation International, this free-to-visit park covering 3342 sq km received official recognition through a pioneering debt-swap agreement with the Bolivian government. Keep your eyes peeled for at least 500 bird species, more than 100 mammals and many, many reptiles, amphibians and insects. *San Borja, Bolivia; entry free.*

Area Protegida Municipal Aquicuana, Bolivia

One of South America's cheapest reserves to reach, this community-run jungle enclave 22km north of Riberalta (B$30 by moto-taxi) is a great spot for birdwatching, wildlife spotting (particularly caimans and anaconda) and fishing for giant *paiche* (a species of bonytongue). *Riberalta, Bolivia; free entry, activity fees vary.*

Lago Tarapoto, Colombia

Lago Tarapoto, 10km west of Puerto Nariño, is a beautiful jungle lake fed by the Amazon, home to pink dolphins, a tiny population of manatees and massive water lilies. It's also easy and cheap to reach – a half-day trip to the lake in a *peque-peque* (a low-slung wooden boat) from Puerto Nariño costs around COP$50,000. *Puerto Nariño, Colombia.*

Jardim Botânico Adolpho Ducke, Brazil

Spanning over 100 sq km, this 'garden' is actually the world's largest urban forest. There's a network of five short trails (guides required, free with admission) and a spectacular 42-meter-high observation tower. The normal entry fee doesn't apply on Tuesdays. *museudaamazonia.org.br; Ave Margarita, Manaus, Brazil; entry R$10-30, free Tue.*

Forte Príncipe da Beira, Brazil

As much an outpost today as when it was founded in the 1770s, star-shaped Forte Príncipe da Beira casts a watchful gaze over the tropical Rio Guaporé. It's an atmospheric ruin today, and combined with the jungle boat trip from Manaus to Porto Velho (R$290, 2½ days) and the shared taxi trip to Guajará-Mirim, it's a real adventure. *Guajará-Mirim, Rondônia, Brazil; free.*

Hiking peaks at São Gabriel da Cachoeira, Brazil

Climbing mountains in the middle of the Amazon is a reason to come to São Gabriel da Cachoeira all by itself. All sorts of jungle excursions are possible too, though you'll pay extra to enter Yanomami tribal lands. Bank on R$250 per day for camping trips and climbs of Serra Bela

Below: the mysterious lands atop Mount Roraima in Venezuela. Right: squirrel monkeys are a common sight in the Amazon canopy.

Adormecida. *São Gabriel da Cachoeira, Rio Negro Basin, Brazil.*

⊕ Waterfall swimming at Taquarussú, Brazil

Waterfalls, caves and pools ring the town of Taquarussú and many are easy to reach on your own, though guides can take you to outlying sites for R$150 per day. Around 1.5km down a well-marked trail, Cachoeira da Roncadeira (70m) and Cachoeira Escorrega Macaco (60m) tumble picturesquely down into small pools good for swimming. *Taquarussú, Tocantins, Brazil; free.*

⊕ Encontro das Águas, Brazil

Most join a pricey tour to view the famous Encontro das Águas (Meeting of the Waters), where the dark Rio Negro pours into the cool, creamy Rio Solimões. An inside tip: the confluence lies on the route of the slow car ferry that joins the two halves of the BR319 highway from Manaus to Porto Velho; passengers are carried for free. *Manaus, Brazil; free.*

⊕ Floresta Nacional do Tapajós, Brazil

If you came to the Amazon to see primary rainforest, look no further than the misty giants of Floresta Nacional do Tapajós. Access is by bus from Santarém, or by boat from Alter do Chão visiting the riverside villages of Maguarí, Jamaraquá and São Domingo, which all have well-maintained forest trails. *Pará, Brazil; entry free, guided hikes and boat trips from R$150 per person.*

⊕ Fordlândia, Brazil

An Amazon excursion with bonus history, the outpost of Fordlândia is what remains of Henry Ford's ill-fated model town, built to source rubber for car parts in 1928 but abandoned in the 1940s. You can wander

cavernous factories, climb the town's water tower and more, at the end of a three- to five-hour boat ride (R$55-70) from Santarém. *Entry free.*

⊕ Canoeing around Mercado de Belén, Peru

At the southeast end of Iquitos, the floating town of Belén consists of scores of rustic huts built on rafts. Seven thousand people live in this frenetic quarter, and canoes float from hut to hut selling and trading jungle produce every morning – get here by taking a cab to 'Los Chinos,' and hiring a canoe-owner to take you around. *Iquitos, Peru; free, charge for canoe transport.*

✋ Budget trekking up Mount Roraima, Venezuela

The inspiration for The Lost World, this dramatic 2810m-high *tepui* (limestone plateau) towers into the clouds. To reach it on a budget, bring your own camping gear and organise a guide from the Pemón community (compulsory) in Santa Elena de Uairén, San Francisco de Yuruaní or Paraitepui. *Paraitepui, Bolívar, Venezuela; entry free, guide fees US$10 to US$20 per day.*

© ©Image Source | Getty Images; Valmol48 | Getty Images

⊙ JUNGLE CRUISES BY PUBLIC BOAT

Boat tours on the Amazon and its myriad tributaries can cost hundreds of dollars, but public river ferries offer rainforest adventures for a fraction of the price. The top tropical river-ride is the four-day, R$350 trip from Manaus to Belém, sleeping in a swaying hammock suspended on deck (bring mosquito repellent or the bugs will eat you alive). For travellers with stamina, the seven-day trip from Manaus up the Rio Solimões to Tabatinga is another jungle epic.

01

CHEAP GALÁPAGOS, ECUADOR

The Galápagos Islands are never cheap, but there are ways to keep a lid on costs. Flights run to Santa Cruz and San Cristóbal, where you can find independent accommodation and arrange short day-tours. Free or cheap wildlife encounters include snorkelling with turtles and sea lions, iguana spotting on the rocks and birding on the tropical shore. *Costs vary.*

02

MEETING PINK RIVER DOLPHINS, BOLIVIA

Freshwater dolphins can be spotted in tropical rivers all over South America, so any boat ride is a photo opportunity. For guaranteed (or almost guaranteed) sightings, head to Trinidad in Bolivia, where boat trips into the Área Protegida Municipal Ibare-Mamoré (per person from B$195) have a very high success rate.

03

BIG RODENTS IN RIO, BRAZIL

The world's biggest rodent, the capybara is found all across South America, but the easiest spots for sightings are the lakes and lagoons near Rio de Janeiro in Brazil. Try the Parque Ecológico Chico Mendes (Barra da Tijuca; free entry) or the Lagoa Rodrigo de Freitas (free entry) in the Jardim Botânico neighbourhood. *Free.*

04

SEE PENGUINS AT MONUMENTO NATURAL ISLOTES DE PUÑIHUIL, CHILE

Boat trips to see Humboldt and Magellanic penguins leave from dozens of spots in southern Argentina and Chile, but you can see these colourful characters for free near Ancud. Buses can get you within 2km of the beach at Monumento Natural Islotes de Puñihuil, where breeding penguins gather from September to March. *Isola Chiloé, Chile; free.*

05

SLOW ENCOUNTERS AT RESERVA NACIONAL TAMBOPATA, PERU

Somnolent sloths are tricky to see in the wild, with green algae growing amongst their fur as natural camouflage. One top spot to seek these slow critters is Peru's Reserva Nacional Tambopata (near Puerto Maldonado, Peru; entry per day S30), which also has clay licks that attract huge flocks of macaws.

06

JAGUARS IN THE PANTANAL, BRAZIL

Seeing the elusive jaguar takes a lot of luck and a fair few bucks. Your best bet is in the Pantanal wetlands in Brazil, on a boat tour from Cuiabá (from R$600 per day), with very strong chances of sightings in the Porto Jofre region. Otherwise, try your luck (best at dawn or dusk) in any of the Amazon's nature reserves.

07

CONDORS IN COLCA, PERU

The plunging valleys of Peru's Cañón del Colca are prime places to spot the magnificent condor, the world's largest flying bird, spiralling the thermals high overhead. From viewpoints on the walking trails around Arequipa, you may even get to see a condor at eye level as it soars majestically past like a kite on the breeze. *Arequipa, Peru; free.*

08

MARINE MAMMAL WATCHING FROM PUERTO MADRYN, ARGENTINA

Península Valdés is Argentina's top spot for whale-watching, and pricey boat tours from Puerto Madryn provide close encounters. However, you can see whales from shore at Observatorio Punta Flecha, sea-lions at the Reserva Faunística Punta Loma and elephant seals at Punta Ninfas, all for free.

SOUTH AMERICA

SOUTH AMERICA'S
WILDEST
EXPERIENCES

Many of South America's most unique critters live in remote quarters, but you don't have to take an expensive tour to meet the locals.

04

08

PATAGONIA

Cutting a dramatic profile across southern Chile and Argentina, Patagonia is South America's great empty quarter. Many of its national parks charge fees, but a rented vehicle opens up epic free hikes to glaciers, mountains, pristine lakes and condor country.

ⓦ Climb Volcán Chaitén, Chile

The centrepiece for the free-to-visit Parque Nacional Pumalín, this rugged volcano blew its top in 2008. Today, hikers make the three-hour round-trip trek to view the eerie, smoking crater with its leafless forests of eruption-charred trees, starting from just north of Chaitén. *conaf.cl/parques/parque-nacional-pumalin-douglas-tompkins; Chaitén, Chile; entry free.*

ⓦ Walking and riding around Palena, Chile

A quiet cowboy town on a turquoise river, Palena's draw is exploring its verdant valleys on foot or horseback. The tourist office can arrange inexpensive guides and horses; consider the five-day trek to remote Lago Palena via charming Rincón de la Nieve farm in Valle Azul, where you can help the family owners round up cattle. *rincondelanieve@hotmail.com; Palena, Chile; homestay incl breakfast CH$18,000.*

ⓦ Reserva Los Huemules, Argentina

Dotted with sustainable homes, private Reserva Los Huemules has 25km of marked trails, offering a quieter and cheaper alternative to Parque Nacional Los Glaciares. Look out for torrent ducks, Magellanic woodpeckers, condors, red foxes and pumas. *loshuemules.com; El Chaltén, Argentina; entry AR$200.*

ⓦ Reserva Nacional Lago Jeinimeni, Chile

Turquoise lakes and the rusted hues of the steppe mark the rarely visited Reserva Nacional Lago Jeinimeni, 52km southwest of Chile Chico. Its unusual wonders range from cave paintings to flamingos, reached via epic hikes in lonely, isolated country – well worth the admission fee. *Chile Choco, Chile; entry CH$3000.*

ⓦ Swim in Parque Nacional Nahuel Huapi, Argentina

A not-too-painful park entry fee provides access to gorgeous bodies of water such as Lago Nahuel Huapi and Laguna Tonchek, which run deep and cold but are great for swims surrounded by elemental scenery. *argentina.gob.ar/parquesnacionales/nahuelhuapi; San Carlos de Bariloche, Argentina; entry AR$300.*

ⓦ Low-cost adventures in Parque Nacional Patagonia, Argentina

Argentina's newest national park is a top spot for birding and hiking, and it's only just being developed. Camp cheaply at El Sauco, or Portal La Ascensión at Lago

Below: the Parque Nacional Torres del Paine in Chile. Right: looking up at the park's not-so-little Perito Moreno glacier.

Buenos Aires, start of a 25km hiking trail that climbs up the *meseta* (plateau). *conaf.cl/parques/parque-nacional-patagonia/; Los Antiguos, Argentina; entry*

Parque Nacional Los Alerces, Argentina

This collection of mountain creeks, mirror lakes and forests of 3600-year-old alerce trees is the unadulterated Andes. Many visit on boat trips from Puerto Chucao, but hiking is free once you pay the moderate entry fee, and there's plenty of cheap camping nearby. *argentina.gob.ar/parquesnacionales/losalerces; Esquel, Argentina; park entry AR$300.*

Hiking from El Chaltén, Argentina

The gateway to Parque Nacional Los Glaciares (North) is an amazing hub for free, self-guided day-hikes, as a warm up for more ambitious multi-day treks into the park. Top destinations in this land of glaciers and sawtooth peaks include short hikes to Chorrillo Del Salto and Los Cóndores, the 18km trek to Laguna Torre and the 20km round-trip to Laguna de Los Tres. *losglaciares.com/en/parque; El Chaltén, Argentina; day hikes free.*

Tramp to Cerro Cristal, Argentina

Glaciar Perito Moreno in Parque Nacional Los Glaciares (South): 30km long, 5km wide and 60m high, and creeping forward at up to 2m per day, causing building-sized icebergs to calve from its face. Most visit on costly cruises from El Calafate, but make your own way to the park by car and the free hike to Cerro Cristal offers stunning views. *losglaciares.com/en/parque; El Calafate, Argentina; entry AR$600.*

Parque Nacional Torres del Paine, Chile

The sheer granite pillars of Torres del Paine dominate the landscape of South America's finest national park. There's a significant entry fee, but come off season and rates are cut in half, meaning discount access to azure lakes, emerald forests, roaring rivers and one big, radiant blue glacier. *parquetorresdelpaine.cl; Puerto Natales, Chile; entry CH$25,000, reduced rates May-Sep.*

Monumentos históricos Nacionales Puerto Hambre & Fuerte Bulnes, Chile

For a manageable fee, you get two national monuments for the price of one at Parque del Estrecho de Magallanes, plus bonus hiking trails. Explore the reconstructed wooden fort of Fuerte Bulnes, then ponder the fate of Puerto Hambre, whose unfortunate inhabitants starved. *parquedelestrecho.cl; Punta Arenas, Chile; adult/child CH$16,000.*

DRIVE THE CARRETERA AUSTRAL, CHILE

Ranked among the world's ultimate road trips, the Carretera Austral runs for 1240km across Chile from Puerto Montt to Villa O'Higgins, passing ancient forests, glaciers, pioneer farmsteads, turquoise rivers and the crashing Pacific. Some drive, some cycle, but all are wowed by the scenery. In fact, highway is a glorified name – the southern half of the route is unpaved and can be in poor condition, and infrastructure is limited, but the adventures are priceless.

SOUTH AMERICA

BRAZIL'S ATLANTIC COAST

Rio de Janeiro is just the front door to Brazil's beautiful Atlantic coast, backed by mountain forests and tinkling waterfalls. To enjoy its thrills on a budget, bring your own board to surf spectacular beaches, or your trekking boots to reach serene forest reserves.

🖐 Make a splash at Praia Itaúna

Praia Itaúna is Saquarema's most beautiful beach and one of the best surf spots in Brazil. National and international competitions are held here each year between May and October. Even if you don't surf, it's a gorgeous spot to swim and bask, overlooked by a church-topped promontory. *Saquarema; free.*

🖐 Natural watersliding at Cachoeira do Escorrega

At the far western end of a forested valley, 2km west of Maromba on the border with Parque Nacional do Itatiaia, an overlapping shelf of rock creates a whooshing natural waterslide dumping swimmers into a gorgeous natural pool. It's a lovely hike with a great reward at the end. *Visconde de Mauá; free.*

🖐 Parque Nacional do Itatiaia

Brazil's oldest national park shelters dense rainforests and rugged upland peaks. Head for the accessible lower section, where short trails lead through lush greenery teeming with monkeys and tropical birds, to waterfalls and natural swimming holes – well worth the entry fee. *icmbio.gov.br/parnaitatiaia; per day foreigners R$33.*

🖐 Hike to pristine sands at Praia Lopes Mendes

This seemingly endless beach with powerful surfing waves (board rentals available) is considered by some to be the most beautiful in Brazil. It's accessible by a 6km trek from Abraão's town beach, or you can travel part way by boat. *Ilha Grande; free.*

🖐 Trindade beaches

About 25km south of Paraty, Trindade occupies a long sweep of stunningly beautiful coastline. Lounge or hike along four of Brazil's most dazzling beaches (Cepilho, Ranchos, Meio and Cachadaço), with surging breakers, vast expanses of white sand, and a calm-water natural swimming pool opposite the furthest beach. *Paraty; free.*

🖐 Parque Nacional Serra dos Órgãos

Created in 1939, this high-country park is best known for its spectacular mountain scenery, superb climbing opportunities and extensive trail system, including the classic 42km, three-day traverse over the mountains from Petrópolis to Teresópolis. *icmbio.gov.br/parnaserradosorgaos; admission lower section/upper section R$32/52.*

THE ATACAMA-SECHURA DESERT STRIP

The deserts of southern Peru and northern Chile couldn't differ more from the Amazon. Bucket-list experiences here, such as ballooning over the Nazca lines, cost big bucks, but cheaper alternatives exist, including great trekking, cycling, surfing and sandboarding.

ⓦ Monumento Natural La Portada, Antofagasta, Chile

Local buses run from Antofagasta to the shore 22km north of the city, where there are stunning views towards this enormous offshore outcrop, eroded into a natural arch by the stormy Pacific. It's the centrepiece of a 31-hectare protected zone, in an area of wild beaches. *Antofagasta, Chile; free.*

ⓦ Cycling around San Pedro de Atacama, Chile

Pedalling around San Pedro de Atacama is like visiting an alien planet. The Valle de la Muerte (Death Valley) has otherworldly red outcrops, or there's hauntingly beautiful Valle de la Luna (Moon Valley), 13km from town. *San Pedro de Atacama, Chile; free, bike rental per day from CH$6000.*

ⓦ Hit Arica's Beaches, Chile

Surfers, swimmers and sunbathers find their niche at Arica. Sunbathers favour Playa El Laucho and Playa La Lisera, south of downtown, but surfers head to Playa Las Machas, a few kilometres north. *Arica, Chile; free, board rental from CH$5000.*

ⓦ Go sandboarding, Peru

The Peruvian town of Huacachina is the place to try sandboarding: rent boards in town and haul them up the slopes of the giant dunes of this desert oasis, for a high-speed rush downhill towards the lagoon. *Huacachina, Peru; board rental from S5.*

ⓦ Hike the highest dune, Peru

About 14km east of Nazca, the wind-sculpted wave of Cerro Blanco is one monster sand dune, 1176m from base to summit. It's a steep three-hour climb to the top, but it's free if you come with your own transport. *Hwy 30A, Nazca, Peru. Free.*

ⓦ The Cañón del Colca, Peru

Arequipa is the starting point for treks in the Cañón del Colca and Cañón del Cotahuasi. Self-sufficient hikers can walk the two- to three-day El Clásico trek through the lower Colca Canyon, or the two-day trip to the source of the Amazon. *Arequipa, Peru; free.*

Below: Cerro Blanco is the largest sand dune on the South American continent.

SOUTH AMERICA

© Lucas Cometto | Getty Images

WELLNESS

Like almost everywhere, South America has caught the wellness bug. Traveller centres such as Peru's Sacred Valley and Chile's Valle del Elqui overflow with yoga centres and sacred sites that – according to believers – vibrate with natural energies. Then there's shamanism; common in indigenous communities, though finding authentic experiences can be a challenge.

At the other end of the spectrum are South America's spas, found all along the coast, and in traveller hubs, where wellbeing typically comes with

Above: the natural world in South America offers great opportunites to experience awe and practise mindfulness.

© Scott Biales | Shutterstock

scented candles and an eyebrow-raising pricetag. Look for smaller, independent operators and natural alternatives to organised wellness activities to make your money stretch further.

SHAMANISM, MASSAGE AND YOGA

South America has its own wellness traditions that predate the arrival of Europeans, but outsiders only started taking notice of the region's shamanistic herbal medicine thanks to William Burroughs and Harvard ethno-botanist Richard Evans Schultes and their 1960s experiments with ayahuasca and other psychoactive plants.

Today, a whole industry has grown up in centres such as Cuzco, Iquitos and Santo Domingo de los Colorados putting travellers in touch with their spiritual side through guided hallucinogenic experiences. Not all are legit, and some are just money-making schemes; considering the risks, many prefer to encounter shamanism through cheaper, less immersive rituals such as coca-leaf tea readings and shamanic massage.

Yoga and other more familiar wellness activities can be found wherever travellers gather, particularly in coastal surf centres and holistic hubs such as the Sacred Valley. Look for drop-in rates at classes run by traveller restaurants for the cheapest sessions.

• **Get good advice** – if you decide to follow the ayahuasca path (and we can't entirely recommend it) seek advice about reputable practitioners from your accommodation and fellow travellers.

• **Massage without the rub** – massages tend to be more expensive in spas and less reputable on the street; strike a happy medium by arranging treatments through budget hotels and hostels.

• **Big city wellness** – urban wellness doesn't just mean yoga classes. In Rio de Janeiro, seek out practitioners of Macumba, a religion that fuses ideas from Catholicism, African religions and indigenous spiritualism. Leaving offerings to the sea goddess Lemanjá on Copacabana Beach is an easy way to dip in a toe.

NATURAL WELLNESS

Natural wellness is easy to find in South America thanks to glorious nature. Swimming holes open up in the jungle, hot springs bubble in forest clearings, glacial lakes bask in icy silence and desert outcrops hum with spiritual energy. There's rarely a charge to dive in, though some springs have been developed as spa resorts, so you may need to seek local advice to find free-to-use hot pools.

• **Head to the desert** – silence, space and spirituality come free once you get away from people: the coastal deserts of Peru and Chile and the high-altitude deserts of the Andes are prime spots for DIY mind-expansion.

• **Find your own falls** – the skyscraper-high cascades often have skyscraper-high entry fees; look for wayside falls on wilderness drives where you can bathe crowd- and cost-free.

• **Break the ice** – watching an iceberg calve off a glacier is just one of many elemental, meditative experiences you can experience for free in the empty wilds of Patagonia.

HIGHLIGHTS

CONSULT A SHAMAN IN CUZCO, PERU

Peruvian shamanism has a powerful call for travellers with a spiritual bent. With some groundwork to find a reputable healer, you can experience all sorts of ritual arts in the Sacred Valley, from coca-leaf tea readings and therapeutic massages to guided hallucinogenic experiences. *p260*

HEALING ENCOUNTERS IN COCHIGUAZ, CHILE

The valley of Cochiguaz is abuzz with new-age energies. How deep you immerse yourself is up to you. Join a yoga class, meditate with only cacti for company, or bathe in a river that cleans the soul; you don't have to believe to enjoy the experience. *p261*

SOUTH AMERICA

CUZCO & THE SACRED VALLEY, PERU

Thousands use Cuzco as a springboard for the Inca Trail to Machu Picchu, but when you need to recuperate at the end of the hike, head to nearby hot springs, book a session with a shaman or take a healing swim in a forest glade.

✋ Aguas Calientes de Marangani

Twenty minutes past Sicuani, this complex of five fabulously hot thermal pools, linked by rustic bridges over unfenced, boiling tributaries, is quite a sight. Join locals washing themselves, their kids and their clothes in the pools for an accessible yet off-the-beaten-track experience. *Occobamba; entry S10.*

✋ Consult a shaman

The psychedelic properties of ayahuasca attract legions of psychonauts, but there's more to Peruvian shamanism than potentially risky drug rituals. Traditional *curanderos* (shaman) offer healing rituals, coca-leaf tea readings and therapeutic massages. Ask at guesthouses for a reputable shaman offering consultations. *Prices vary.*

✋ Make the Q'oyoriti pilgrimage

Sacred Ausangate mountain (6384m) is the *pakarina* (mythical place of sacred origin) of llamas and alpacas, and the frozen purgatory for condemned souls. Pilgrims gather for the festival of Q'oyoriti in May/June for paganistic dances and whipping by men in white masks representing *ukukus* (mountain spirits). The three-hour trek to the site starts at Mahuayani. *Free.*

✋ Pre- and post-trek soaking at Santa Teresa

Trekking to Machu Picchu is cheaper from Santa Teresa than from Cuzco. An added bonus is the restorative soaking offered at the Baños Termales Cocalmayo, a series of landscaped hot springs, with a natural shower straight out of a jungle fantasy and, as an extra perk, beer and snacks. *Cocalmayo; entry S10.*

✋ Jungle swims at Quillabamba

Wild swimming at Quillabamba is another recommended tonic for trek-tired muscles. La Balsa, hidden far down a dirt track, is a bend in the Río Urubamba that's perfect for swimming and river tubing, while Mandor, Siete Tinajas and Pacchac are beautiful waterfalls where you can splash, climb and eat jungle fruit straight off the tree. *Quillabamba; free.*

✋ A room with a massage in Cuzco

In a characterful old building close to La Catedral, Amaru is deservedly popular. It is certainly the only Cuzco hotel with a complimentary massage: a welcome treat after tackling the Inca Trail. *amaruhostal.com; Cuesta San Blas 541; single/double from S182/231.*

VALLE DEL ELQUI, CHILE

The South American version of Shangri-La, Valle del Elqui is Chile's capital of new age and alternative living. Practise yoga beside desert streams, meditate on mountains and get lost in dark skies: this a place where mindfulness counts more than money.

✋ Look to the (dark) skies

The almost rain-free Elqui Valley has emerged as one of South America's top destinations for watching dark skies – something guaranteed to make you consider your place in the universe. The hordes head to the observatory at Mamalluca; for free star-gazing, try the fields and rocky hills immediately around the village of Pisco Elqui. *Free.*

✋ Chilean Buddhism?

Centro Otzer Ling is a Buddhist stupa unexpectedly transported to the Andes – you'll think you took a wrong turn off the Pan-American Highway and somehow ended up in the Himalaya. Complete with wafting Tibetan prayer flags, this snow-white monument is now a popular spot for meditation against a parched, mountainous backdrop. *otzerling.com; Estupa Cochiguaz, Cochiguaz; free.*

✋ Esoteric energy at Cochiguaz

The new-age capital of northern Chile, the secluded side-valley of Cochiguaz is credited with an extraordinary concentration of cosmic vibes. As well as formal centres for yoga and meditation, plenty of visitors find inner peace for free, meditating in the cactus-dotted mountains and bathing in the gently splashing Elqui river. *Cochiguaz; free.*

✋ Veggie treats and yoga at Govinda's

In an airy space across from Vicuña's plaza, Govinda's offers an alternative spin on the Andes, with a lip-smacking menu of vegan and vegetarian food, healthy juices and yoga classes that fill the restaurant with mats several days a week. *C/Arturo Prat 234, Vicuña; meals CH$2500-3500.*

✋ Secret springs

The mountains around Vicuña are dotted with natural springs, both hot and cold, where you can bathe in front of breathtaking mountain scenery. You'll need local help to find many of these al fresco spas – ask for directions to the Termas Las Hediondas, where warm waters are channelled into pools. *Free.*

Below: the Milky Way over the Valle del Elqui in Chile, where the stars seem sharper than anywhere else.

SOUTH AMERICA

THE BEST THINGS IN LIFE IN OCEANIA: TOOLKIT

Every landscape, from deserts to rainforests, can be found somewhere in Australia, New Zealand and the Pacific, and much of this nature is on display for free. To get the best out of this natural wonderland, seek out free wilderness experiences and bargain street food.

Bargains pop up everywhere in Oceania, from some of the world's best free-to-visit national parks to gratis museums in every major city. However, don't be fooled into thinking a trip here will be cheap (get that out of your system with an Asia stopover en route).

Australia and New Zealand are high-wage economies, and the bills for meals and organised activities can be eye-watering; make use of campsites and self-catering facilities to make your dollars stretch further.

TOP BUDGET DESTINATIONS IN OCEANIA

Exploring this staggering collection of islands can easily rack up a giant bill; squeeze more out of your bucks in the following hubs:

• **QUEENSLAND, AUSTRALIA:** Fewer big cities, more big nature, and plenty of bargains.

• **ADELAIDE, AUSTRALIA:** Manageable living costs and easy access to South Australia's winelands.

• **QUEENSTOWN, NZ:** Great low-cost living if you can resist the urge to splash out on activities.

• **NORTHLAND, NZ:** The north end of New Zealand; missed by most, meaning fewer people to push up prices.

• **FIJI & TAHITI:** The best beaches for the fewest bucks in expensive-to-explore Polynesia.

Below: both Adelaide and Melbourne have tram networks and easy-to-use travel cards for speedy urban travel. Check zones and times for cheaper fares.

© amophoto_au | Shutterstock

TRANSPORT

You haven't experienced long journeys until you've travelled in Oceania. It can take a week to drive comfortably from Perth to Sydney, or Auckland to Invercargill, and island-hopping in the Pacific adds time and cost. With these kinds of distances, getting anywhere costs money. Public transport fizzles out quickly in the interior; a rental car or camper can take a lot of the pain out of crossing from coast to coast.

AIR TRAVEL

Flights across Oz and NZ are a cheap deal per kilometre. There are a few budget airlines – Jetstar Airways (jetstar.com) is a major player – but they only offer minor savings on the main carriers. On hops between Australia and New Zealand, Scoot (flyscoot.com) has cheap flights from second-tier airports.

• **Get a pass** – the Qantas Walkabout Airpass (qantas.com) adds six reduced-cost domestic flights (plus New Zealand) to your international flight, but you need to book all the legs at the same time.

• **Go all the way round** – round-the-world and 'circle Pacific' fares offer savings if you plan to visit Australia, New Zealand and the Pacific. OneWorld (oneworld.com) and Star Alliance (staralliance.com) are the major players.

• **Skip Christmas** – demand and prices go mad around Christmas; give it a month or so either side and costs come way down.

TRAIN

Australia and New Zealand have limited rail networks. A handful of state railway companies run cost-effective local services connecting cities to their hinterlands. Long-haul services run by Journey Beyond (journeybeyondrail.com.

au) and Great Journeys of New Zealand (greatjourneysofnz.co.nz) are more like luxury tours.

BUS

Across Oceania, buses go everywhere travellers want to go, but ye gods can they take a long time. Fares are less painful – this is easily the cheapest way to travel long-distance. In Australia, Greyhound Australia (greyhound.com.au) is head honcho; InterCity (intercity.co.nz) zips all over New Zealand.

• **Act your age** – most bus companies offer generous discounts for seniors, students and children.

• **Go by night** – it's rarely the most comfortable night's sleep, but often unavoidable with the distances involved, and you'll save the price of a night's accommodation.

CAR & CAMPER

Now we're talking! Australia and New Zealand were made for road trips, with epic distances, amazing scenery, and cheap camping available almost

Above: in Western Australia, the 4WD is king, here in Francois Peron National Park.

everywhere. There's a well-oiled machine hiring cars and campervans to travellers, with plenty of scope for one-way hire. Make sure you're covered for dirt roads if you plan to access campgrounds, and stick to a 4WD for serious off-road tracks.

• **Check the small print** – unlimited mileage is what you're after; otherwise it's easy to run up a whopping bill for the extra kilometres.

• **Rule the roads** – the RV/camper/motorhome is king, and rates are reasonable once you factor in the savings on accommodation.

• **Avoid the toll** – both Australia and New Zealand have toll roads that you can easily bypass on free routes.

• **Be breakdown safe** – carry food, plenty of water and blankets, and keep your phone charged; on back-country roads, it could be days before help comes along.

• **Learn the road rules** – watch for giant trucks and suicidal wildlife (roos are a major hazard after dark).

LOCAL TRANSPORT

Commuter rail lines can be a bargain traveller's best friend, providing easy access from major cities to national parks, trailheads and ferry ports. In most cities, a pre-paid card covers all forms of transport, offering savings on pay-as-you-go fares.

• **Ride free** – free public transport is taking off in city centres: Perth, Melbourne and Sydney all have free routes.

• **Rideshare** – try Zoomy and Uber in New Zealand; Uber, Ola and DiDi in Australia. Also look out for more informal car-pooling apps like Coseats (coseats.com).

• **Pedal power** – bike hire is available all over the place, or grab a cheap secondhand bike from an op shop (charity/thrift shop); helmets are mandatory.

Above: renting or buying and then selling a campervan is a popular way of getting around the sights of Australia and New Zealand.

OCEANIA

ACCOMMODATION

Accommodation will bite a big chunk out of your bank balance, but travellers who don't mind a shared dorm or canvas (or an RV roof) overhead will be in budget heaven. It's not quite so peachy in the Pacific; hostels are rare, campsites are rarer, and most beds are in beach resorts, though family-run pensions are becoming more common.

CAMPING

Despite the abundance of wilderness, you can't camp just anywhere in Australia and New Zealand. Many areas are off-limits (with fines to back this up), but national park campsites are a steal, and you can 'freedom camp' for nothing at many designated rest areas.

Aussie and Kiwi campsites are a dream: trees for shade, loos, sometimes solar-powered showers, free barbecues and power hook-ups, and wildlife on tap (the challenge is keeping the local fauna out).

• **Car camp** – with a tent and sleeping mat, any vehicle becomes a mobile home. Carry water, an esky (icebox) and camping stove (or use site barbecues) to keep a lid on eating costs.

• **Use the app** – to find free camping sites, try Wikicamps (wikicamps.com.au), Travellers Autobahn Road Trip (travellers-autobarn.com.au) or Hipcamp Australia (hipcamp.com), which covers camping spots on private farms.

• **Hit the Huts** – in New Zealand, national park huts dot the main walking trails, offering basic beds for affordable prices. Rates range from nothing up to NZ$140; camping outside the huts is often free.

• **Just ask** – in the Pacific, camping is usually possible in the grounds of hotels, resorts and pensions; ask locally to see who will let you set up a tent for a fee.

HOSTELS, HOTELS & GUESTHOUSES

Australia and New Zealand have a staggering number of hostel options, from rustic rooms behind bush pubs to party-all-night backpacker resorts where getting a good night's sleep is rarely the objective. Find beds through HostelWorld (hostelworld.com) or the Youth Hostel Association (yha.com.au in Oz; yha.co.nz in NZ).

The next cheapest beds are at guesthouses, B&Bs and pubs with noisy, rough-and-ready rooms above the bar; prices creep up at weekends; for cheap choices, try Airbnb (airbnb.com.au) and Hosted Accommodation Australia (australianbedandbreakfast.com.au). Farm Helpers in NZ (fhinz.co.nz) can put you in touch with farms that offer a bed for free in exchange for farm work; in Oz, try farmstaycampingaustralia. com.au.

• **Plan ahead** – for the best bargains, find a bed before you arrive. If you want to wing it, backpacker lodges send staff with a van to round up customers at bus stands.

• **Homestay** – a chance to interact with local people and sample home cooking, often for less than a hotel bed; try homestay.com for listings.

• **Holiday homes** – holiday lets ('bachs' or 'cribs' in New Zealand) can be a cheap bet in boondocks locations, or for long stays on the coast; check with local letting agents.

• **Pacific-specific accommodation** – in Samoa, look out for *fale* (simple thatched beach huts, sometimes with blinds for walls); in PNG and Melanesia, Christian missions often offer basic accommodation; you could also negotiate a bed and a meal directly with villagers.

PRO TIPS

Across Oceania, in Australia, New Zealand or amongst the Pacific nations, one of your biggest costs will be transport. Distances are vast on land – and not small between islands – so whether you're flying, boating or driving, moving from place to place is rarely cheap. Once you've arrived though, camping (or a basic cabin or beach hut) is a low-cost option, and sometimes free. And food is often inexpensive for the quality you'll be eating – be it a freshly caught crayfish, juicy seasonal mangoes, or a full Fijian *lovo* (a feast cooked in an underground oven). Expect sumptuous and memorable meals that defy your cost expectations.

Tasmin Waby, writer and editor

OCEANIA

ARTS & CULTURE

Australia and New Zealand have imported a taste of the old world to the new world, with top-notch galleries and museums – many free to visit – stacked with old masters and ancient treasures. Refreshingly, the cultural gaze also takes in indigenous creativity, with growing numbers of exhibitions focusing on local culture.

In the Pacific Islands, you'll find fewer oil paintings and more ethnological artefacts from cultures clinging onto their traditional way of life. Take advantage of opportunities to meet with

Above: Downtown Melbourne from the Southbank, where the National Gallery of Victoria and the Southbank Theatre are based.

indigenous people and learn about their legends, customs and crafts; there's usually a small cost, but it's a great opportunity to support a threatened way of life.

OCEANIA'S TOP SWITCHEROOS

If you plan to notch up the top must-sees, you'll need a stack of cash to smooth the way; save by swapping top sights for some of the following cheaper, more intimate experiences.

• **SKIP:** Waitangi Treaty Grounds, Paihia, New Zealand (entry NZ$50)

• **SEE:** Te Papa Museum, Wellington, New Zealand (entry free)

• **SKIP:** Melbourne Museum, Melbourne, Australia (entry A$15)

• **SEE:** Australian Museum, Sydney, Australia (entry free)

• **SKIP:** Sydney Opera House, Sydney, Australia (seats from A$70)

• **SEE:** Southbank Theatre, Melbourne, Australia (seats from A$35)

• **SKIP:** Museum of Old & New Art, Hobart, Tasmania (entry A$30)

• **SEE:** National Gallery of Victoria, Melbourne, Australia (entry free)

• **SKIP:** Aboriginal Art Sites, Kakadu National Park, NT, Australia (park pass A$40)

• **SEE:** Split Rock Aboriginal Art Site, Queensland, Australia (entry free)

OCEANIA'S CULTURE FOR LESS

There's loads of free culture in Oceania's big cities, from free museums and galleries to live music and stand-up comedy in city pubs (where the only cost is a schooner of beer) to street art (almost everywhere).

• **Art is education** – in most big cities, major art galleries and museums don't charge entry, part of an ethos of making culture free for the people.

• **Civic pride** – the state or national museum is almost always free; it's a badge of honour for state capitals in the competition for cultural status.

• **Hit the library** – state libraries are often packed with free exhibitions and oddball displays such as Ned Kelly's armour in Melbourne (p271).

• **Check out the town hall** – most state and national parliament buildings offer free tours on weekdays; call ahead to register.

• **Flash your ID** – International Student Identity Card (ISIC; isic.org) is good for discount entry fees and great savings on transport. Seniors get good discounts too.

• **Grab a pass** – some cities have pre-paid tourist cards offering promo prices for tours and mainstream visitor attractions.

• **Snap the big things** – one of Australia's most entertaining free activities is snapping photos of the 'big things' (the Big Banana, the Big Prawn, the Big Merino Sheep... the list goes on) as you drive around this vast nation.

• **Swap tall towers for lookouts** – Australia has the most skyscrapers per head of population in the world, but skyscraper observation decks cost megabucks; there's almost always a natural viewpoint nearby offering views that are almost as good and cost nothing.

• **Sacred sites** – Oceania's churches are mainly Johnny-come-latelies, but many ancient indigenous sites are free to visit; hire a local guide for insights into what these spaces mean to their traditional custodians.

HIGHLIGHTS

ART GALLERY OF NEW SOUTH WALES, SYDNEY, AUSTRALIA

A grand neoclassical facade hides a stunning collection of works by local talent (Sidney Nolan, Grace Cossington Smith, John Olsen and more) as well as high-impact imports. Perhaps the most impressive feature is the Yiribana Gallery, with amazing creative output from Aboriginal and Torres Strait Islands communities. *p268*

TE PAPA, WELLINGTON, NEW ZEALAND

Te Papa roughly translates as 'treasure box' and New Zealand's national museum certainly earns that name, with a stupendous collection of Māori artefacts and its own *marae* (traditional meeting house) given as much status as the displays on European settlement history. *p272*

OCEANIA

SYDNEY, AUSTRALIA

Australia's de facto capital – in the eyes of its residents, anyway – Sydney is a city of iconic buildings, set on one of the world's greatest harbours. It's bold, brash and beautiful, but big-city living can leave budgets wrecked in its path, unless...

Art Gallery of New South Wales

Behind a neoclassical facade, this fantastic gallery exhibits works by international artists as well as locals, including Sidney Nolan, Grace Cossington Smith, John Olsen and Arthur Boyd. The Yiribana Gallery is a treasure-house of Aboriginal and Torres Strait Islanders' art. *artgallery.nsw.gov.au; Art Gallery Rd, The Domain; free.*

Conservatorium of Music

Wander around the halls of this international music school for free, or book a paid tour to hear about the building's past life, then wrap your ears around some world-class classical music at one of many free monthly recitals and concerts. *sydney.edu.au/music; 1 Conservatorium Rd; closed Sun; free.*

Government House

Set within the Royal Botanic Garden, this Gothic sandstone mansion (built 1837–43) is the official residence of the Governor of NSW. Its lovely loggia looks over a formal garden but the interior can only be accessed on a free guided tour; collect your ticket from the gatehouse. You'll need ID. *governor.nsw.gov.au; Macquarie St; free.*

Museum of Contemporary Art

Sitting on the iconic waterfront overlooking Circular Quay, this enormous Art Deco building is a cathedral of creativity. Entry to the main galleries is free and you can also join complimentary tours, including some conducted by teenaged guides, to gain a youthful insight into modern art. *mca.com.au; 140 George St, The Rocks; closed Mon; free.*

Sydney Opera House

Opera House architect Jørn Utzon stormed out before his magnum opus was fully complete, but you can view his finished creation, and for for free. If your budget doesn't stretch to a

Below: the city itself is a living work of art thanks to its natural setting and some judicious architectural commissions.

OCEANIA

performance, pop inside for a nibble in one of the restaurants – Opera Kitchen (operabar.com.au) is the least expensive. *sydneyoperahouse.com; Bennelong Point; admission free.*

The Rocks
Sydney's oldest quarter is where the First Fleet landed in 1788, an event regretted by the indigenous population. Once a den of iniquity, the cobbled streets are now lined with food markets, Aboriginal art galleries and cafes. The Rocks Discovery Museum (2-8 Kendall Lane; free) explains how these historic buildings were saved. *therocks.com; Kendall Lane; Fri-Sun; free.*

White Rabbit Gallery
Housed in what once served as a Rolls-Royce depot, this eccentric gallery showcases Chinese contemporary art. The 1400 works, by more than 500 artists, are rotated regularly, and there's also a peaceful teahouse serving Taiwanese teas and Chinese *cha. whiterabbitcollection. org; 30 Balfour St, Chippendale; closed Mon & Tue; free.*

Outdoor Cinema
Big screens appear everywhere during summer, including a pricey screen outside the Opera House. Olympic Park in the west, however, screens new, cult and classic flicks for free every summer; you can bring a picnic, and it's easy to reach via public transport. *sydneyolympicpark. com.au; Olympic Park; free.*

Centennial Park
Reminiscent of London's great green spaces, this 130-year-old city park is home to lakes, an intricate labyrinth, more than 15,000 trees, free barbecues, and a colony of flying foxes which you can meet during the monthly Bat Chat. Come to

people-watch (and bat-watch) and feel a bit like a local. *centennialparklands. au; Oxford St; free.*

Ocean pools
A saltwater swim is a rite of passage if you want to plug into Sydney's outdoors culture. Some pools, including Bondi's Icebergs, charge a fee, but most are free. Check out Giles Baths (a natural rockpool or 'bogey hole'), McIvers Baths (women only), Ross Jones Memorial Pool and Maroubra's spectacular Mahon Pool. *Free.*

Royal Botanic Garden
In big, brash Sydney, this historic, free-to-access park is an oasis of tranquillity. Take a free tour or just explore the 200-year-old collection of plants at your own pace. Don't miss the 'Cadi Jam Ora – First Encounters' section, honouring the Cadigal, Sydney's original inhabitants, and their relationship with the land. *rbgsyd.nsw.gov.au; free.*

Sydney Harbour Bridge
The giant 'coat hanger' no longer dominates the Sydney skyline, but it stands as one of the twin icons of the Harbour City, along with the Opera House. An official climb to the top of the bridge is expensive – a budget alternative is to wander the pedestrian pathway along the bridge's eastern edge, from where the views are still stunning. *Free.*

Sydney Harbour National Park
This 4-sq-km NP protects Sydney's foreshore and its heritage. There's no charge to visit historic landmarks such as the Quarantine Station (qstation.com. au; 1 North Head Scenic Dr) and the Fort Denison Martello tower on Pinchgut Island, although you need to buy ferry tickets to get here. *nationalparks.nsw.gov. au; free for pedestrians.*

CAMPING FOR URBAN EXPLORERS

After exploring Sydney's cultural riches, you can hit the outback without leaving the city limits at Royal National Park (royalnationalpark. com.au; Audley Rd, Audley), the planet's second-oldest national park. As well as roaming around bush-covered and ocean-stroked wilderness on 100km of walking trails, you can wild camp at Uloola Falls and North Era for just A$12 (possibly Sydney's cheapest accommodation). Visit via public transport (train to Waterfall, or ferry from Cronulla) for free admission.

OCEANIA

MELBOURNE, AUSTRALIA

Lacking Sydney's instant sex appeal, Melbourne seduces inquisitive visitors more subtly, revealing its manifold charms with the more time you spend here. Whether you also spend a ton of money depends on what you know and where you go.

✋ Arty alleyways

Melbourne's labyrinthine laneways are lined with coffee houses, bars and the colourful work of guerrilla graffiti artists. You can pay to do a tour, but it's easy to freestyle it, starting from Hosier Lane and weaving north(ish). Besides local talent, many international artists have contributed, including Banksy, Blek le Rat and Shepard Fairey. *24hr; free.*

✋ Australian Centre for Contemporary Art (ACCA)

In a rusty angular building that evokes the site's industrial past, this contemporary art gallery exhibits challenging work by Australian and international artists. Out front, you can't miss *Vault*, a controversial sculpture by Ron Robertson-Swann, dubbed the 'Yellow Peril' and 'Steelhenge' by critics. *accaonline.org.au; 111 Sturt St; closed Mon; free.*

✋ Koorie Heritage Trust

Melbourne is a sprightly young 185-years-old, but this centre addresses the neglected history of those who occupied this land for the preceding 30,000-plus years, displaying artefacts, staging exhibitions and explaining the oral history of the southeastern Aboriginal peoples. You can join tours (per person A$30) along the Yarra. *koorieheritagetrust.com;*

Yarra Bldg, Federation Square; free.

✋ National Gallery of Victoria (NGV)

The always-free NGV has a permanent collection of old and contemporary work by international superstars from Constable to Rodin, spread over two sites. The St Kilda Rd premises is stunning, with a moat out front and stained-glass atrium inside. Nearby, the Ian Potter Centre (Federation Square; free) houses the Australian collection, including mind-blowing Aboriginal art. *ngv.vic.gov.au; 180 St Kilda Rd; free.*

Below: street art in Melbourne ranges from massive murals to fast-moving throw-ups in the city centre's Hosier Lane.

⛭ State Library of Victoria

From the funky pavement sculpture to the epic domed reading room, this is a house of stories about Victoria's past, present and future, with books, exhibits (including Ned Kelly's armour) and talks/debates – some held in the Wheeler Centre (established by LP's founder; wheelercentre.com; 176 Little Lonsdale St) and many free. *slv.vic.gov.au; 328 Swanston St; free.*

⛭ Australian Centre for the Moving Image (ACMI)

An Aussie-accented celebration of movies, television and dynamic digital culture, ACMI is a cave of wonders for film fans. The star is the interactive Screen Worlds exhibit, which explores the evolution of moving art from zoetrope to modern game labs. You can watch free programmes from the National Film and Sound Archive in the Mediatheque. *acmi. net.au; Federation Sq; free entry.*

⛭ Music for nada

Live comedy and music is in Melbourne's blood, and there's plenty of both at venues throughout the city and in the suburbs – often with no entry charge. St Kilda's Espy (espy.com.au; 11 The Esplanade) is an iconic live-band boozer, where front-bar gigs are typically free. In town, Cherry Bar (cherrybar.com.au; 68 Little Collins St) puts on the type of acts suited to its previous rock and roll address (AC/DC Lane). Listings appear in free street-press publications such as Beat (beat.com.au).

⛭ Abbotsford Convent

This former convent, dating back to 1863, is a rambling collection of ecclesiastical architecture that's home to a thriving arts community of galleries, studios and cafes. Free tours of the complex run on Sundays, or download the Abbotsford Convent app for a self-guided walking tour created by Wurundjeri elders, musicians and artists. *abbotsfordconvent.com.au; 1 St Heliers St, Abbotsford; free.*

⛭ Shrine of Remembrance

One of Melbourne's icons, the Shrine of Remembrance is a commanding memorial to Victorians who have served in war and peacekeeping. Built between 1928 and 1934 with Depression-relief, or 'susso' (sustenance) labour, its design is partly based on the Mausoleum of Halicarnassus. There are panoramic views of Melbourne's skyline from its terraces. *shrine.org.au; Birdwood Ave, South Yarra; free.*

⛭ Tour Parliament

The grand steps of Victoria's parliament (1856) are often crowded with tulle-wearing brides or placard-holding protesters, but on sitting days the public is welcome inside to view proceedings from the galleries. On non-sitting days there are eight guided tours daily; times are posted online and on a sign by the door. Numbers are limited to 25 people, so arrive at least 15 minutes before time. *parliament.vic.gov.au; Spring St; closed Sat & Sun; free.*

⛭ Free laughs

Free stand-up comedy is easy to find in Melbourne. On Mondays, the gratis laughs are at Spleen Bar (comedyatspleen. com; 41 Bourke St, CBD). On Tuesdays, the free-giggles action shifts to the Jazz Rooms at the Lido Cinema (facebook. com/lidocomedytuesday; 675 Glenferrie Road, Hawthorn), while Thursdays feature regular free shows at the Rochester Hotel (rochey.com.au; 202 Johnston St, Fitzroy).

THRIFTY COMMUTING

Exploring Melbourne for next to nothing is simple, thanks to an easy-to-navigate street grid and several initiatives. The Central Business District (CBD) is a 'free zone', meaning you can ride trams here gratis (ptv.vic.gov. au). Alternatively, board the City Circle free tourist tram (yarratrams. com.au), which trundles around town with running commentary. Alternatively, enjoy a free walking tour led by volunteers from the City of Melbourne (international-greeter.org/destinations/melbourne).

OCEANIA

WELLINGTON, NEW ZEALAND

The huge number of museums, theatres, galleries and arts organisations is completely disproportionate to the size of this pocket-friendly NZ capital. Wellingtonians are rightly proud of their city, so grab a craft beer or gourmet coffee and find out why.

Te Papa
There's a clue in the name – Te Papa roughly translates as 'treasure box', and New Zealand's national museum is certainly home to many riches: an amazing collection of Māori artefacts including a colourful *marae* (meeting house); natural history and environment exhibitions; Pacific and NZ history galleries; and the National Art Collection. *tepapa.govt.nz; 55 Cable St; free.*

Wellington Museum
For an introduction to Wellington's social and maritime history, head to this bewitching little museum inside an 1892 bond store on the wharf. Highlights include a moving documentary on the sinking of the Wahine ferry, and Māori legends dramatically told using tiny holographic actors. *museumswellington. org.nz; 3 Jervois Quay, Queens Wharf; free.*

A cable-car museum by cable car
It's worth paying the modest fare on the historic Wellington Cable Car (wellingtoncablecar.co.nz; NZ$5/9 one-way/return) to visit the dainty Cable Car Museum (museumswellington.org. nz; 1a Upland Rd, Kelburn; free), which tells its story. Ramble back to town

through the Wellington Botanic Garden *(wellingtongardens.nz; 101 Glenmore St, Thorndon; free).*

New Zealand Parliament
New Zealand has one of the oldest continuously functioning parliaments in the world: it was the first to give women the vote (in 1893) and the first to include an openly transgender member (in 1999). Learn about NZ's unique version of democracy on a free guided tour. *parliament.nz; Molesworth St; free.*

City Gallery Wellington
Housed in the monumental old library (built 1893), Wellington's much-loved City Gallery pulls in acclaimed contemporary international exhibitions, as well as unearthing up-and-coming new talent from the NZ scene. Charges apply for major temporary exhibits. *citygallery.org. nz; Civic Sq; free.*

Hit the trails
Urban art trails, we mean. Start with the waterfront Wellington Writers Walk (wellingtonwriterswalk.co.nz), with passages by NZ writers executed in stone typography, then hit the free Wellington City Walk (sculpture.org.nz/walks/ wellington-city-walk), which takes in 17 quirky public sculptures. *Free.*

OCEANIA

FIJI

Fiji's fascinating culture is a medley of traditions and beliefs from India, Europe and the Pacific, best experienced for free in the small but cosmopolitan capital, Suva, population 93,970 and the mightiest metropolis in the South Pacific.

✋ Fiji Museum

The modest entry fee to Fuji's national museum buys access to a host of riches. The centrepiece is Fiji's last *waqa tabus* (double-hulled canoe). Other exhibits include the shoe of consumed missionary Thomas Baker and the rudder from the *Bounty*. *fijimuseum.org.fj; Ratu Cakobau Rd, Suva; closed Sun; entry F$7.*

✋ Mariamma Temple

Suva's technicolour, gopuram-topped Hindu temple is legendary for the South Indian fire-walking rituals that take place here during July or August. *Howell Rd, Samabula, Suva; free.*

✋ Parliament of Fiji

Indigenous Fijian values are apparent throughout the art-filled national parliament buildings. The *vale ne bose lawa* (parliament house) takes its form from the traditional *vale* (family home) and has ceremonial access from Ratu Sukuna Rd. Free tours are offered if you email in advance. *parliament. gov.fj; Battery Rd, Suva; Mon-Fri; free.*

✋ Tavuni Hill Fort

This fort was built by Tongan chief Maile Latumai in the 18th century. There are grave sites, a *rara* (ceremonial ground) and a *vatu ni bokola* (head-chopping stone). *Sigatoka, Viti Levu; entry F$12.*

✋ Naiserelagi Catholic Mission

About 25km southeast of Rakiraki, the Naiserelagi mission church is famous for its mural depicting a black Christ, in a bark-cloth *sulu* (sarong) with a *tanoa* (wooden bowl) at his feet; Fijians make offerings of mats, *tabua* (whale's tooth), flowers and oxen. *King's Rd, Rakiraki, Viti Levu; entry free, donation requested.*

✋ Wasavula Ceremonial Site

At this macabre pre-colonial site, there's a sacred monolith that villagers believe grew from the ground and an altar for cannibalistic ceremonies, with a *vatu ni bokola*, another rock for the severed head and a bowl-like stone where the brain was placed for the chief. *Vanua Levi; free.*

Below: Fiji's parliament building in Suva features traditional architecture and a national collection of art.

OCEANIA

© Getty images | iStockphoto

TOP CULTURAL ENCOUNTERS

Indigenous people across Oceania are inviting visitors to see the new world from their perspective, minus the colonial gaze. Consider the following cost-friendly cultural encounters.

BARUNGA FESTIVAL, KATHERINE, AUSTRALIA

For three days over a wonderful long weekend in mid-June, Barunga, a bushland outpost 80km east of Katherine, is the setting for a massive celebration of indigenous arts and crafts, dancing, music and sporting competitions. There's a fee, but it includes camping (bring your own equipment). *barunga festival.com.au; Katherine, Northern Territory, Australia; per person A$55.*

NAVALA VILLAGE, FIJI

Navala's chief enforces strict town-planning rules: dozens of traditional thatched *bure* (local houses) are laid out neatly in avenues, with a central promenade sloping down the banks of the Ba River. It's a photographer's delight, but you need to get permission and pay the entrance fee. *Navala, Nausori Highlands, Viti Levu, Fiji; entry F$25.*

QUINKAN COUNTRY ROCK ART, QUEENSLAND, AUSTRALIA

The lonely Split Rock Gallery near Laura is the only Quinkan Aboriginal rock-art site open to the public without a guide. The sandstone escarpments here are covered with paintings thought to date back 14,000 years; if you're here by yourself, it's both eerie and breathtaking. *Peninsula Developmental Rd, Laura, Queensland, Australia; donations appreciated.*

INJALAK, ARNHEM LAND, AUSTRALIA

Injalak Hill (Long Tom Dreaming) is one of western Arnhem Land's best collections of rock art, including a rare depiction of Yingarna, the female Creation Ancestor. Visit on well-worth-the-money tours from Injalak Arts led by indigenous guides (the centre is free to visit). *injalak.com; Gunbalanya (Oenpelli), Arnhem Land, Australia; tours A$110.*

TARI MARKET, PAPUA NEW GUINEA

Secluded in the highlands, Tari is one of the few towns in PNG where you may spot people still wearing traditional dress – the distinctive clothing of 'wigmen' from the Huli tribe. The main market days are Monday, Wednesday and Friday and you may meet locals in their finery, though it's a fading custom. *Tari, Hela Province, Papua New Guinea; free entry.*

WAITANGI TREATY GROUNDS, PAIHIA, NEW ZEALAND

On the site where Māori chiefs signed the Treaty of Waitangi, this historic site has an entry fee that covers a guided tour, spirited performances of Māori rituals and dances, and the treasure-packed Te Rau Aroha and Museum of Waitangi (which offers a frank discussion of the inequities of the treaty). *waitangi.org.nz; 1 Tau Henare Dr, Waitangi, New Zealand; entry NZ$50.*

BRAMBUK CULTURAL CENTRE, HALLS GAP, AUSTRALIA

Run by five Koorie communities (including the traditional custodians of the region) in conjunction with Parks Victoria, this excellent, family-friendly centre near Melbourne offers insights into local culture and history (A$5). At the time of writing, it was closed for renovation; check before visiting. *277 Grampians Rd, Halls Gap, Australia; by donation*

FOOTPRINTS WAIPOUA, NEW ZEALAND

Led by Māori guides, this four-hour twilight tour into Waipoua Forest is a fantastic introduction to both Māori culture and the forest giants. It's an organised event, with a fee, but it's still a moving and mesmerising encounter. *footprintswaipoua.co.nz; Copthorne, 334 SH12, New Zealand; per person NZ$105.*

© Mark Read | Lonely Planet

FOOD & DRINK

Australia and New Zealand are as foodie as they come – almost to the point of vanity when it comes to local wine, coffee and craft beers. Plugging into the local food scene can be as expensive or as economical as you like, depending on where you eat, and how much formality you require from your dining environment.

A white tablecloth and fine wine by the bottle is always going to cost more than a budget steak and stubby (375ml beer bottle) at a Returned Servicemen's Club. Across the region, cheap street food and canteen meals at shared tables can

Above: save money by buying fresh produce when it is local and in season at South Melbourne. Right: you'll find the best coffee in the world down under.

be every bit as tasty as five-star dining, and there's always the coin-operated (sometimes free) public barbecues as a fallback position.

Different rules apply in the Pacific Islands, where the most interesting, economical dining is found in tiny island capitals, catering to office workers and civil servants, while travellers are often hostage to the menus and prices in their resort. To find the best gourmet grub on a budget, head to the local market or business district on a weekday lunchtime and see what locals are chowing down on.

SELF-CATERING & STREET FOOD

With abundant hostel and campground kitchens, self-catering is an easy way to save. Stock up on fresh local ingredients at farmers' markets and take full advantage of the free or coin-operated barbecues in public parks and national-park campgrounds. With an inexpensive fishing license, you can add freshly caught fish to the menu (try beach fishing at sunset with pipi shells as bait).

Street food is Oceania's other big expense-deterrent. At markets, beachfronts, harbours and disused downtown parking lots you'll find everything from the comforting (meat pies, chiko chicken spring rolls) to the cool (fusion burgers, noodle stir-fries) and the cordon bleu (artisan cheeses, oysters and mussels fresh off the seabed).

On the islands of the Pacific, resort accommodation can distance you from the local cuisine and the food prices paid by locals. Head to the island capitals for the best portable feasts: stalls selling barbecued meat and seafood, breads and buns, and things steamed in leaves pop up at markets, transport hubs and boat jetties.

• **Cook on the move** – with a camping stove or a camper-van kitchen, you can sleep cheap and eat cheaper. Carry an esky (cool box) to keep things fresher for longer.

• **Be a barbecue buddy** – snags (sausages), steaks and even fresh bugs (slipper lobsters) can be picked up cheaply at butchers, fish markets and harbour kiosks and taste better sizzling off the barbie.

• **Court the food court** – most malls have good, inexpensive Asian-style food courts, serving everything from roast beef sandwiches to laksa soup and sushi.

• **Eat on the quay** – wherever fish are landed, you'll find cheaper prices on seafood, from freshly shucked oysters to fish and chips.

• **Time it right** – farmers' markets tend to run from early morning to lunchtime, while stalls targeting city workers are busy downtown at lunchtime.

SIT-DOWN DINING

Oceania has both bargain cafes and fine-dining menus that will make your

CAFE CULTURE

Don't neglect your coffee habit. Aussies and Kiwis insist their coffee is the best in the world (to the chagrin of Italians) and coffee-shops, kiosks and mobile coffee vans everywhere can knock you up a perfect flat white for a stimulating price, often with an inexpensive breakfast or a lunchtime snack alongside. Seek out single-origin coffees and local bean producers such as Sydney-based Coffee Alchemy (coffeealchemy. com.au) in Australia, and Wellington's Havana Coffee Works (havana. co.nz). Learn the local lingo: a flat white is a stronger latte; a long black is a strong Americano; and a ristretto is a strong espresso shot.

OCEANIA

eyes water; just don't expect the kind of dining-for-loose-change you may have encountered if you stopped over in Asia on the way here. A typical main course in Australia will cost north of A$15, with New Zealand prices close behind. On Pacific islands, many locals opt for takeaways over expensive sit-down restaurant meals, particularly in the evening when prices almost double.

Across Oceania, the cheap lunch is king: midday specials abound downtown, and you can always fall back on street-food hubs set up where office workers gather if none of the lunch deals tickle your fancy. Fiji has the added advantage of Indian fast food – curries, kebabs and rice dishes served canteen-style at bargain prices.

- **Tipping tips** – tipping isn't as vigorously applied in Australia and New Zealand as it is in the US or Europe. Service charges are rarely added to the bill, except on holidays; at other times, tipping is optional and many people just leave loose change.
- **Local life-savers** – in Australia, look for super-cheap (but not too adventurous) meals at the local Returned Servicemen's Club (rslnational.org) or Surf Lifesaving Club (sls.com.au).
- **Follow the local lead** – student quarters, arty and migrant neighbourhoods and areas around big-city markets are a rich vein when it comes to finding budget restaurants.
- **Pub meals rock** – in Australian Outback towns, the local pub is often the town restaurant, shop and fuel station rolled into one, good for cheap, filling meals (if not for variety).
- **Take it away** – restaurants across Oceania offer take-outs at a big discount; take dinner to the nearest beach for a sunset view and you won't mind the lack of table service.

HIGHLIGHTS

CARRIAGEWORKS FARMERS MARKET, SYDNEY, AUSTRALIA

The king of Sydney's farmers' markets, with producers, sellers and food stalls clustered in the heritage-listed Carriageworks gallery. Grab a flat white from an artisan coffee stand and browse, sampling tasty morsels as you go. *p280*

GET SHUCKED, BRUNY ISLAND, TASMANIA, AUSTRALIA

Freshly harvested oysters are shucked to order and served up with wasabi sour cream and optional sparkling wine at this oyster-farm kiosk overlooking Great Bay. It's agreeably informal, appealingly economical and a top spot for budget epicureans on the move. *p283*

JEANNERET WINES, CLARE VALLEY, AUSTRALIA

Ravishing rosés are the hightlight at Jeanneret, tucked away on a dirt track in the Clare wine country. Inexpensive tastings (redeemable against purchases) don't just cover wines; you can also sample the beers made on site by the Clare Valley Brewing Co. *p285*

YEALANDS ESTATE, MARLBOROUGH REGION, NEW ZEALAND

Wine-making is taken to the 21st century (and simultaneously back a century or two) at this zero-carbon winemaker near Blenheim, where sheep keep down the grasses around vines that stretch endlessly around a futuristic hub. Sample the end product for free at the cellar door. *p284*

SMART DRINKING

Australia and New Zealand are top producers of craft beer and fine wine, but those chocolatey notes and summer fruits come at a cost, particularly at the artisan end of the market. The good news is that even the cheap wine is pretty darned good, and tasting fees at most vineyards are not much more than the cost of a single glass at bar prices (and can usually be redeemed against purchases).

Wherever you go, bars and clubs will hit your wallet harder than pubs and restaurants. Camping types often skip formal drinking entirely in favour of a slab (multipack) of bottled beer or wine by the goon (box), sipped around the campfire. In the Pacific, beer is cheap, wine expensive, and cocktails either cheap or dear, depending on whether local or imported spirits are used.

Left top: an Asian hawker market, Sydney-style, for cheap eats. Left bottom: Lentil As Anything vegetarian cafe in Abbotsford, Melbourne. Above: a food market in Adelaide, Australia.

OCEANIA

SYDNEY, AUSTRALIA

Sydneysiders are unashamedly food-obsessed: connoisseurs of coffee; aesthetes of real ale; champions of artisan cheese; deli dilettantes. But this doesn't always translate to high prices, particularly in the city's excellent coffeeshop cafes.

OCEANIA

Carriageworks Farmers Market

Over 70 regular stallholders sell their goodies at Sydney's best farmers' market, held in the heritage-listed Carriageworks gallery. Coffee and snack stands do a brisk business and vegetables, fruit, meat, seafood and sunflowers are sold in a convivial atmosphere. *carriageworks. com.au; Carriageworks, 245 Wilson St, Eveleigh; Sat.*

Pablo & Rusty's

Mega-busy and loud, with close-packed tables, this excellent cafe is high energy. Seriously good coffee (several single-origins available daily) is complemented by generously proportioned breakfast and lunch specials, ranging from sourdough sandwiches to wholesome world-food combos. *pabloandrustys. com.au; 161 Castlereagh St, CBD; Mon-Fri; lunches A$9-25.*

Fine Food Store

In the pub-heavy Rocks, it's a delight to find this contemporary cafe that works for a sightseeing stopover or a better, cheaper breakfast than your hotel. Genuinely welcoming staff make very respectable coffee, plus delicious panini, sandwiches, brunches and generously large cocktails. *finefoodstore.com; cnr Mill & Kendall Lanes, the Rocks; light meals A$10-19.*

Gumshara

Prepare to queue for some of Sydney's best broth at this cordial ramen house in a popular but unglamorous Chinatown food court. They boil down over 100kg of pork bones a week to make the gloriously thick stock. There are lots of ramen bowls, including some that pack quite a punch. Ask for extra back fat for a real indulgence. *gumshara.com; shop 211, 25-29 Dixon St, Chinatown; ramen $12-19.*

Sydney Fish Market

This piscatorial precinct on Blackwattle Bay has fishmongers, restaurants, sushi and oyster bars, delis and a cooking school. Chefs, locals and overfed seagulls haggle over mud crabs, lobsters and salmon at the early morning fish auction; ready-to-eat snacks include salt and pepper squid, fish and chips and oysters. *sydneyfishmarket.com.au; Bank St, Pyrmont; dishes from A$10.*

Cow & the Moon

Forget the diet and slink into this cool corner cafe, where an array of sinful truffles and tasty tarts beckons. Resist and head straight for some of the world's best gelato – this humble little place won the 2014 Gelato World Tour title in Rimini, Italy. *cowandthemoon.com.au; 181 Enmore Rd, Enmore; gelato from $6.50.*

Below: enjoying classic (and cheap) fish and chips at Sydney Fish Market. Right: a bargain Aussie pie from Bourke Street Bakery in Surry Hills.

Bourke Street Bakery

Lining up outside this teensy bakery is an essential Surry Hills experience. It sells a tempting selection of pastries, cakes, bread and sandwiches, along with pies and near-legendary sausage rolls. There are a couple of spots to sit inside, but on a fine day you're better off on the street. *bourkestreetbakery.com.au; 633 Bourke St; Surry Hills; baked goods A$5-12.*

Reuben Hills

An industrial design and creative Latin American menu await at Reuben Hills, set in a terraced house and its former garage. Come for great coffee, refreshing homemade *horchata* (rice milk), stellar fried chicken, tacos and *baleadas* (Honduran tortillas). *reubenhills. com.au; 61 Albion St, Surry Hills; mains A$9-22.*

Pilu Baterro

It may be attached to the upmarket Pilu restaurant, but this casual deck cafe is an altogether more relaxed, beachy affair. It's a top spot for Italian-influenced pre-surf breakfasts, delicious post-surf lunches or evening grazing. The coffee is great, but so is the Sardinian wine by the glass. *pilu. com.au; Moore Rd, Freshwater; closed Mon & Tue; dishes A$12-22.*

Black Star Pastry

Wise folks follow the black star to pay homage to excellent coffee, totally brilliant cakes and a few very good savoury things (gourmet pies and the like). Queues form for the famous strawberry-watermelon cake (dubbed 'the world's most Instagrammed cake' by the New York Times), but some say the raspberry-lychee is better. *blackstarpastry.com.au; 277 Australia St, Newtown; snacks from A$6.*

Golden Lotus

Delicious bowls of pho, crunchy textures and fresh flavour bursts make this one of the best of Newtown's crop of vegan and vegetarian restaurants. As well as vegetable-based meals, there are dishes involving soy-based chicken and fish substitutes. It's BYO alcohol. *goldenlotus-vegan.com; 343 King St, Newtown; mains A$14-19.*

Bargain barbecues

Eating out can cost a fortune in Sydney, but eating outside will only set you back the price of a few 'snags' (sausages). Seek out the free or coin-operated barbecues in Western Sydney Parklands (westernsydneyparklands.com.au) and Bronte Park (next to Bronte Beach), where locals and visitors bond over shared, if very well used, hotplates.

Chinatown night markets

Sydney's Chinatown is a pan-Asian melting and cooking pot, where you can tuck into a host of cuisines including Cantonese, Thai, Malaysian, Japanese and Vietnamese. It's all about *yum cha* (eating dim sum and drinking tea) at lunchtime, but go Friday evening for the sensational street-eats market. *Dixon St, Chinatown; admission free.*

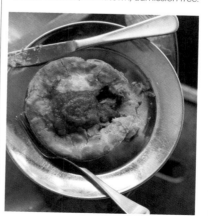

CHEAP GETTING AROUND

Trips from food stop to food stop needn't cost a fortune. On public transport, buy an Opal Card (opal.com.au), and travel costs are capped at A$16.10 per day, or A$50 per week, no matter how many bus, tram, train or ferry trips you take. To save if driving, be aware that fuel is cheaper earlier in the week, and more expensive at the weekend (when prices are hiked to match demand); download the Motormouth app (motormouth.com. au) for the best up-to-date petrol prices.

OCEANIA

AUCKLAND & HAURAKI GULF, NEW ZEALAND

The twin harbours of cultured, civilised Auckland frame a narrow isthmus punctuated by volcanic cones and the islands of the Hauraki Gulf, all studded with foodie gems. You'll eat well here, for not that much, before diving headlong into nearby natural wonders.

🥄 Chuffed, Auckland

Concealed at the rear of a building, this street art-decorated place is a contender for the inner-city's best cafe. Grab a seat on the terrace and tuck into cooked breakfasts, Wagyu burgers or flavour-packed toasted sandwiches. *chuffedcoffee.com; 43 High St, CBD; mains NZ$10-21.*

🥄 Zool Zool, Auckland

This co-production between two of Auckland's top Japanese chefs takes the traditional *izakaya* (Japanese pub) uptown. Enjoy some of the city's best ramen noodle dishes, frosty mugs of beer, tempura squid, soft-shell crab and panko-crumbed fried chicken. *zoolzool.co.nz; 405 Mt Eden Rd, Mt Eden; closed Sun & Mon; dishes NZ$10-19.*

🥄 Azabu, Auckland

Splash some cash for the blend of Japanese and Peruvian influences at Azabu. Standout dishes include the tuna sashimi tostada and Japanese tacos with wasabi avocado. *azabuponsonby.co.nz; 26 Ponsonby Rd, Grey Lynn; mains & shared plates NZ$16-39.*

🥄 Giapo, Auckland

Expect elaborate constructions of ice-cream art topped with all manner of goodies, as Giapo's extreme culinary creativity and experimentation produces possibly the planet's best ice-cream extravaganzas. *giapo.com; 12 Gore St, CBD; Fri-Sun; ice cream from NZ$12.*

🥄 Swallow, Great Barrier Island

Hands-down the best burgers on Great Barrier Island, within easy strolling distance of the sands at Kaitoke Beach. Massive buns are stuffed with pork belly or beef and blue cheese. *facebook.com/BurgerShackGBI; Main Rd, Claris; burgers A$15-17.*

🥄 Murray, Piha

Cool beats, coffee and soft-shell-crab tacos are dispensed from this surf shack a short walk from the beach at Piha. Seafood fans should ask if mussel fritters or raw fish salad are available. *facebook.com/murray.inc; Marine Pde, Piha; Fri-Sun; mains from NZ$10.*

Below: Asian flavours and expertise have infused Antipodean cooking, such as here at Zool Zool in Auckland.

© Brett Atkinson | Lonely Planet

OCEANIA

SOUTHERN TASMANIA, AUSTRALIA

Hobart is the hospitable hub for a reasonably priced eating expedition through the flavours of Tasmania. When you tire of the city, continue the gut-busting at Eaglehawk Neck, Port Arthur and south along the coast at Bruny Island and the Huon River estuary.

Salamanca Market, Hobart

Open-air Salamanca Market fills an eponymous tree-lined square with stalls touting everything from fresh organic produce to crafts, antiques and quality street food. Grab breakfast, a bagel or burger at the fantastically popular Retro Cafe (31 Salamanca Pl; mains $11-20) and enjoy the people-watching show. *salamancamarket.com.au; Salamanca Pl, Hobart; Sat.*

Flippers, Hobart

Out on Constitution Dock, the shack housing Flippers has a voluptuous fish-shaped profile. Fillets of flathead and curls of calamari come straight from the deep blue sea and into the deep fryer. *flippersfishandchips.com.au; 1 Constitution Wharf; fish & chips $12-17.*

Jackman & McRoss, Hobart

Make sure you stop by this enduring Hobart favourite, even if it's just to gawk at the display cabinet full of delectable pies, tarts, baguettes and pastries. Breakfasts include scrambled egg, bacon and avocado panini or potato, asparagus and brie frittatas. *facebook. com/Jackman-and-McRoss; 57-59 Hampden Rd, Battery Point; breakfasts and snacks $6-15.*

Cubed, Eaglehawk Neck

Drink in the views as well as the coffee at this restored, solar-powered caravan at the Pirates Bay Lookout. Blankets and cushions are spread out at the roadside, looking out at the famed Totem Pole and Candlestick across the bay. *cubedespresso.com.au; Pirates Bay Dr, Eaglehawk Neck; Fri-Mon; snacks $3.50-5.*

Get Shucked, Bruny Island

Celebrating the 'fuel for love', this small producer has a tasting room and wooden deck overlooking Great Bay. Wolf down a briny dozen with wasabi sour cream and a cold flute of Jansz sparkling wine. Shucking brilliant. *getshucked.com.au; 1735 Bruny Island Main Rd, Bruny Island; six oysters unshucked/shucked from $12/15.*

Masaaki's Sushi, Geeveston

What a surprise – Tasmania's best sushi is in sleepy Geeveston! Opening hours are limited (and they usually sell out by 2pm), so book a table ahead or join the long queue for takeaway. *20b Church St, Geeveston; Fri-Sun; 6/12 portion sushi plates $9/18.*

OCEANIA

MARLBOROUGH REGION, NEW ZEALAND

The northeastern tip of NZ's South Island has the perfect wine-growing climate for crisp Sauvignon Blancs. You can drink like a king here, for a king's ransom, or slip under the radar at small wineries and save some cents for a bottle to take home.

🖐 Mills Bay Mussels, Havelock

How would you like your green-lipped mussels? Steamed with white wine and garlic? Grilled in garlic butter? Beer-battered? In a chowder? Try them all at this seaside shack right by the marina, with a shared table inside and more outside. *millsbaymussels.co.nz; 23a Inglis St, Havelock; Thu-Sun; mains NZ$6.50-17.*

🖐 Scotch Wine Bar, Blenheim

A versatile and sociable spot, Scotch offers local wines, craft beer on tap and delicious sharing plates for moderate prices. The southern fried chicken with chili mayo is a crispy and filling crowd-pleaser, but consider the raw hangar steak with juniper, pine needles and sorrel. *scotchbar.co.nz; 24-26 Maxwell Rd, Blenheim; Tue-Sat; dishes NZ$7-26.*

🖐 The Burleigh, Blenheim

The humble pie rises to stratospheric heights at this fabulous bakery-deli on Blenheim's rural fringes. Try the pork-belly or steak and blue cheese, or perhaps both; just avoid the lunchtime rush. *facebook.com/theburleighnz; 34 New Renwick Rd, Blenheim; Tue-Sat; pies NZ$6-9.*

🖐 Brancott Estate, Fairhall

Marlborough's most impressive cellar door is poised atop a hillock overlooking the vines. Visitors with deeper pockets come for fine dining and vineyard tours, but the standard tastings are quite reasonably priced. *brancottestate.com; 180 Brancott Rd, Fairhall; tastings NZ$5-15, redeemable against purchases.*

🖐 Allan Scott Family Winemakers, Rapaura

This esteemed vineyard is in an appealing location, with modernist timber, concrete and tin buildings set amid the vines; the wines are great, and the attached bistro (mains NZ$10-17) is good, too. *allanscott.com; 229 Jacksons Rd, Rapaura; Wed-Sun; tastings NZ$3, free with purchase or lunch.*

🖐 Yealands Estate, Seddon

Experience free tastings at this zero-carbon winery near Seddon. A self-drive vineyard tour includes spectacular views over Cook Strait, and glimpses of babydoll sheep grazing between the vines (they keep the grass down, but can't reach the grapes). *yealands.co.nz; cnr Seaview & Reserve Rds, Seddon; Thu-Mon; free.*

OCEANIA

SOUTH AUSTRALIAN WINELANDS

Adelaide is the portal through which you enter a world of sophisticated wineries, rolling vineyards and elegant eateries that couldn't be further from Australia's Outback image. Pick your tastings and there's no need to bust your budget while you quaff.

Central Market, Adelaide

This place is an exercise in sensory bombardment: a barrage of smells, colours and cacophonous stallholders selling gourmet produce. Cafes, food courts and Adelaide's Chinatown are here too. Do not miss. *adelaidecentralmarket.com.au; Gouger St; closed Mon; snacks from A$5.*

Peel Street, Adelaide

Peel St is Adelaide's after-dark epicentre, lined with hip drinking and eating options, including this place. You'll pay a mark-up but it's worth it to join glam urbanites sipping South Australian wine. *peelst.com.au; 9 Peel St; Wed-Sat; mains from A$20.*

Barossa Farmers Market, Angaston

Every Saturday, the farm shed behind Vintners Bar & Grill fills with stalls selling farm-fresh produce and local vintages at fair prices. Expect hearty Germanic offerings, coffee and questionable buskers. *barossafarmersmarket.com; 740 Stockwell Rd, Angaston; Sat.*

Yalumba, Angaston

Yalumba is one of the Barossa's major producers of budget wines, but there's nothing 'budget' about their gorgeous 1850s estate. The cellar door offers tastes of the good stuff that doesn't end up in cardboard wineboxes. *yalumba.com; 40*

Eden Valley Rd, Angaston; tastings from A$10, redeemable against purchases.

Mr Mick, Clare Valley

Swing by Mick's winery/restaurant for a tasting, or tapas-style plates ($12-25), with wines by the glass (we rate the tempranillo). *mrmick.com.au; 7 Dominic St; Clare; tasting A$10, redeemable against purchases.*

Jeanneret Wines

For some lovely rosé, seek out Jeanneret Wines down a dirt road. Sip a few glasses at the bar, then repair to the deck with a BYO picnic. Clare Valley Brewing Co (cvbc. beer) is here too. *jeanneretwines.com; 22 Jeanneret Rd, Sevenhill; tasting A$10, redeemable against purchases.*

Below: Adelaide's Central Market is one of the southern hemisphere's best food and drink experiences, with plenty of tasty titbits to sample.

OCEANIA

FESTIVALS & EVENTS

Oceania's festival calendar is 40% cultural, 20% musical, 20% sporty, 10% foodie and 10% out-there eccentric – dry boat regattas in the desert, tossing the tuna, war canoe races, you name it. One thing common to all Oceania festivals is a squeeze on accommodation and transport; book far in advance to beat the rush for the limited number of cheap seats.

PRACTICALITIES

Big festivals are never truly cheap in Oceania; already expensive cities

Above: in sports-mad Melbourne, the MCG is the top venue for cricket and Australian Rules football matches, but you can catch local matches anywhere.

© Neale Cousland | Shutterstock

become agonisingly overpriced, and demand for beds and tickets vastly outstrips supply. Don't lose faith, however: take a longer trip overlapping festival dates and prices drop to more manageable levels.

• **Plan a year ahead** – to get to Australia or New Zealand at a reasonable price for Christmas, start shopping before the turkey has gone cold the year before.

• **Early bird specials** – many paid-for festivals offer a limited number of cut-price tickets for people who book early; check event websites for release dates.

• **Be non-traditional** – when there's a crush, you may have better luck finding a cheap bed at couchsurfing.com.

• **Camping** – represents your best chance of a cheap stay, though there's heavy demand for campsites too; a relative with a big garden can come in very handy!

RELIGIOUS FESTIVALS

Christian holidays are celebrated with gusto in Australia and New Zealand, if you can get over the quirk of celebrating Christmas in the blistering heat of summer; while Hindu, Muslim and animist celebrations are big in the Pacific (and great for cheap eats). For Christmas/New Year, even coach-class seats go for almost business-class prices; stay from November to February and you might have a shot at a cheap seat.

MUSIC & THE ARTS

Australia and New Zealand rock. Events like Ultra Australia (ultraaustralia.com; Sydney, Melbourne; March) and Rhythm & Vines (rhythmandvines.co.nz; Gisborne, New Zealand; December) feature the same big stars and ticket prices found everywhere; camping can reduce costs but it's never going to be a cheap day (or days) out. For pocket-pleasing culture, look for free concerts and performances at city-organised cultural and arts festivals.

More and more events are celebrating indigenous culture, from Aboriginal *corroborees* (traditional gatherings) to canoe races across the Pacific. Most charge entry fees, but few mind paying to support marginalised communities.

FOOD

Food festivals such as Gourmet Escape (gourmetescape.com.au; Margaret River, Australia; November) and Marlborough Wine & Food Festival (marlboroughwinefestival.com; Brancott Estate, New Zealand; February) have big-name guest chefs and winemakers and hard-to-digest ticket prices, though it can be a cheaper-than-normal way to sample treats from five-star cooks. Look for smaller foodie stages such as Whitianga's Scallop Festival (scallopfestival.co.nz; Whitianga, New Zealand; September) to avoid the squeeze.

SPORTS

Aussies, Kiwis and Pacific Islanders are sports mad, and you'll pay top dollar to attend any big rugby, cricket, tennis or Aussie rules football event. Look instead to non-professional sports: surfing competitions such as the Margaret River Pro (margaretriver.com/event/the-margaret-river-pro; Prevelly, Australia; Mar) and the Australian Surf Life Saving Championships (sls.com.au/aussies; Perth, Australia; Apr) are free to watch, and there are modest fees for goofy events like the waterless Henley-on-Todd Regatta (henleyontodd.com.au; Alice Springs; Australia; Aug).

HIGHLIGHTS

SYDNEY GAY AND LESBIAN MARDI GRAS
Perhaps the world's liveliest gay and lesbian get-together, and a serious party for people of all persuasions, Mardi Gras fills the streets of Sydney with pumping music, fabulous fashion and lots of flirting. If you crack the cheap accommodation problem, there's no charge to join in most of the festivities. *p289.*

PASIFIKA FESTIVAL, AUCKLAND
Celebrating all things Polynesian, this Auckland spectacular brings together every strand of New Zealand's collective identity, from Māori and Pacific Island culture to the newly awakened sense of pride amongst all New Zealanders at the islands' pre-European culture. *p288*

OCEANIA

FESTIVALS

Adelaide Festival, Australia

Adelaide's biggest annual jamboree features top-flight international and Australian dance, drama, opera, literature and theatre performances. Some shows are ticketed but many are free, including public readings by famous writers and sometimes even the main opening extravaganza. *adelaidefestival.com.au; Adelaide, Australia; Mar; some events free.*

Bula Festival, Fiji

Tickets are needed for many events at this Fijian celebration, but funds raised go to local charities so it's worth the small investment. Nadi is the centre of the action, with plenty of dance, music, art and pageants, plus interesting, multicultural food. *facebook.com/bulafestivaltrust; Nadi, Fiji; Jul; some events free.*

Heiva Festival, Tahiti

Tahiti's biggest festival features lots of island culture, from costumed dances (once used as war rituals) to canoe races and coconut-tree-climbing competitions. It's spread over a month, with events taking place around the islands. *heiva.org; Pape'ete, Tahiti; Jun-Jul; many events free.*

Laura Aboriginal Dance Festival, Australia

This sleepy Cape York settlement comes alive in June with the three-day Laura Aboriginal Dance Festival, Australia's largest traditional indigenous gathering. Proceeds from ticket sales support indigenous dance and other artforms. *anggnarra.org. au/our-country/laura-dance-festival; Laura, Australia; Jun or Jul alternate years.*

Milne Bay Kenu & Kundu Festival, Papua New Guinea

This racing gala sees dozens of canoes full of warriors adorned in traditional dress and paddling to the beat of island drums. Alongside, brilliantly attired groups perform dances amid much eating, drinking and revelry. *nationalkenukundufestival.com; Alotau, Milne Bay Province, Papua New Guinea; Nov; admission free.*

Moomba, Melbourne, Australia

The biggest free festival in the city, Moomba attracts more than a million people to the banks of the Yarra for a feast of fireworks, live music and aquatic shenanigans including the Bird Man Rally, where people attempt to 'fly' over the river in homemade contraptions. *moomba.melbourne.vic.gov. au; Melbourne, Australia; Mar; free.*

Pasifika Festival, Auckland, New Zealand

Celebrating the colourful Pacific Islands' culture that makes up so many strands of New Zealand's collective identity, this festival is a fulfilling (and filling) fiesta of food, art, craft and more live music than you can shake a *rakau* (stick) at. *aucklandnz.com; Western Springs, Auckland, New Zealand; Mar; free.*

Perth Festival, Australia

Held over several weeks, WA's biggest cultural event spans paid-for and free theatre, classical music, big-name rock and jazz, visual arts, dance, international films (screened in a beautiful outdoor cinema) and a writers' week. *perthfestival.com.au; Perth, Australia; Feb-Mar; some events free.*

Below: the free Symphony Under the Stars in Adelaide. Right: there are free firework displays in Sydney several times during the year, including New Year's Day and Australia Day.

🖐 Queenstown Winter Festival, New Zealand

NZ's fanciest Alpine resort can get pretty bling during the ski season, but the 10-day Winter Festival offers loads of free events amid all the frivolity, including a firework-laden lake-front party and a free-to-enter 'suitcase race' down the slopes of Coronet Peak. *winterfestival.co.nz; Queenstown, New Zealand; late Jun-early Jul; free.*

🖐 Sculpture by the Sea, Australia

From late October to early November, the cliff-top trail from Bondi Beach to Tamarama transforms into an exquisite sculpture garden. Serious prize money is on offer for the most creative, curious or quizzical creations, while spectators can just stroll and admire. *sculpturebythesea. com; Bondi to Tamarama coastal walk, Sydney, Australia; Oct-Nov; free.*

🖐 Sydney Gay and Lesbian Mardi Gras, Australia

Sydney's free-to-watch Mardi Gras is a riot of colour, music and fun, with outrageously flamboyant costumes and an ocean of rainbow flags. Come early to stake out a spot, or book a table on a restaurant balcony to watch in comfort (for a price). *mardigras.org.au; Oxford St, Sydney, Australia; first Sat in Mar, some events ticketed.*

🖐 Sydney New Year's Eve, Australia

The biggest party of the year, with spectacular firework displays shooting off the Harbour Bridge. There's a variety of regulated zones where you can watch the show, some ticketed but many free, though the best free vantage points fill up early – sometimes even by mid-morning. *Sydney, Australia; 31 Dec-1 Jan.*

🖐 Te Matatini National Kapa Haka Festival, New Zealand

This spine-tingling haka competition occurs in odd-numbered years, with much gesticulation and tongue waggling. It's not just the haka: expect traditional Māori song, dance, storytelling and other performing arts. Main events are ticketed, but prices drop for 'early bird' bookings. *tematatini. co.nz; New Zealand, host cities vary; Mar.*

🖐 Wellington Summer City, New Zealand

So ecstatic are Wellingtonians at the arrival of summer that they erupt into a carousal of concerts, dances and cultural happenings. The windy city's three-month long festival strings together nearly 100 events, most of which are as free as the breeze. *wellington. govt.nz; Jan-Mar; free.*

🖐 Yabun, Sydney, Australia

Indigenous people celebrate Australia Day as 'Survival Day', gathering for free live music, dances on *corroboree* grounds, arts and sports and discussions of the issues facing indigenous communities. It's both invigorating and thought-provoking. *en-gb. facebook.com/YabunFestival; Victoria Park, Sydney, Australia; 26 Jan; free.*

AUSTRALIA'S ODDEST FESTIVALS

Australia loves an oddball celebration. As well as the legendary waterless Henley-on-Todd Regatta near Alice Springs in August, consider the Parkes Elvis Festival, when hundreds of Elvises gather five hours inland from Sydney. (parkeselvis-festival.com.au; Jan), tuna-tossing at Tunarama (tunarama.net; Port Lincoln, South Australia; Jan) and the Deni Ute Muster (deniutemuster. com.au; Deniliquin; Oct) celebrating singlets, superfluous car lights and the humble pick-up truck. You'll pay less for a ticket than at Australia's mainstream fests, meaning a lot more fun for your Aussie dollar.

OCEANIA

OUTDOORS
& ADVENTURE

For many people, Australia and New Zealand are defined by the great outdoors. There are few places in the world with quite so much natural wonder on the doorstep of every city, town and village. Both countries have made big bucks from organised activities – trekking, reef trips, sailing, rafting, skiing, sky-diving, bungie jumping (first developed as a leisure activity in Auckland, New Zealand) – but adventures for free spill out of the landscape almost everywhere too.

The Pacific Islands are another outdoor

Above: New Zealand's Tongariro National Park. Right: keep your eyes peeled for koalas in Australia's treetops.

playground, but they tend to be remote, sometimes right on the edge of the explored map, requiring extra effort (and sometimes cost) to unlock their wonders.

That said, the pristine tropical jungles, encounters with tribal people, rainbow-colored reefs and kilometres of beaches – often described as the best in the world and rarely fenced off or restricted from public use – are nearly always free.

NATIONAL PARKS & NATURE RESERVES

Australia and New Zealand's national parks are the envy of the world. Areas vast beyond measure are set aside as havens for flora and fauna, both on land and offshore and, refreshingly, most are free to enter, though some charge not-too-painful fees for drivers and campers (a good reason to travel by eco-friendly pedal power).

Camp in a national park campsite in Australia and the wildlife will come to you: kangaroos and wallabies mill around wherever there's grass and water, while possums swing down from the trees at night to search for snacks. Alongside national parks, many city parks are just extensions of the wilderness, poking into built-up areas.

On top of officially protected areas, there's just a heck of a lot of wilderness, particularly in New Zealand's mountains, Australia's Red Centre, and the interiors of most of the Pacific Islands. Add in amazing marine life that can be spotted for free offshore and it's an impressive package for thrifty nature-lovers.

AUSTRALIA

In Australia, more than 600 natural areas are designated as national parks, protecting 4% of the land, with another 6% protected as state forests, nature reserves, conservation areas and indigenous protected areas. With the notable exceptions of Uluru-Kata Tjuta National Park and Kakadu National Park, there are few designated entry fees for national parks, though there may be a charge of A\$8-15 to enter by car, motorcycle or camper. State forests, nature reserves, and indigenous areas are often free to visit.

Parks Australia (parksaustralia.gov.au) maintains six of the biggest reserves, but each state has its own Parks & Wildlife department with oversight for local parks and reserves. Most state governments offer National Parks Passes for A\$25/60/120 per week/month/year, covering vehicle access to all parks in the state. Camping is possible in almost every national park for a fee of A\$7 to A\$12 per person – the cheapest accommodation in Australia if you have a tent or camper.

HIGHLIGHTS

CHEAP SNORKELLING ON GREAT KEPPEL ISLAND, AUSTRALIA

Most of the Great Barrier Reef lies far from shore, but at Great Keppel Island, the reef sneaks in close to the beach. Bring your own snorkelling gear, and the only cost is the effort of trekking out to the best snorkelling beaches from the ferry drop-off point at Fisherman's Beach. p294.

TONGARIRO ALPINE CROSS-ING, NEW ZEALAND

An epic day-hike through an other-worldly landscape of steaming vents, bubbling hot springs, psychedelic lakes and towering volcanic craters in Tongariro National Park. You'll need to arrange a drop-off from town, but then the adventure's free, less a bit of strenu-ous effort climbing the ridges. p300.

OCEANIA

NEW ZEALAND & THE PACIFIC

New Zealand's 13 national parks and 10,000 protected areas conserve an astonishing 32% of the islands' landmass, under the watchful oversight of the Department of Conservation (doc.govt.nz). Most national parks are free to visit, but some reserves charge moderate access fees for visitors and vehicles. You can cross many parks on foot on New Zealand's ever popular Great Walks (doc.govt.nz/parks-and-recreation/things-to-do/walking-and-tramping/great-walks).

Camping is available almost everywhere, from free 'freedom camping' to organised sites with water, toilets and barbecues (around NZ$15 per person); some mountain routes have 'huts' charging as little as NZ$5 for rudimentary facilities, or as much as NZ$140 for well-equipped serviced huts on the Great Walks.

Parks and reserves in the Pacific tend to be less organised, though substantial areas of land are protected. Some charge, some don't, some are so wild that there's nobody around to pay even if you wanted to. The biggest cost is normally getting to these remote areas and paying for local guides to show you around.

ADVENTURE ACTIVITIES

Organised activities are the foundation of the tourist industry in Oceania, and locals have pulled out all the stops to help visitors spend their money moving through, under or over the region's ravishing natural environment. Prices, however, can be stratospheric (for multi-day treks, bareboat sailing, jeep tours, skydives and boat trips); or minimal (for walking day-tours); or might be nothing at all (for some trips led by national park staff).

• **Support small operators** – big operators with shiny vehicles charge shiny prices; small independents offer a more intimate experience, lower prices and better chances of meeting the wildlife.

• **Consider distance** – reef trips cost less where the reef comes close to shore; treks cost less starting from the trailheads rather than downtown.

• **Day trip or stop over?** – overnight trips cost much more; if you can go and come back on the same day, costs tumble.

• **Carry basic kit** – your own wetsuit, mask and fins will open up all sorts of DIY adventures, from reef snorkelling to canyoning. All are available locally at cheap prices.

• **Hit the 'op shop'** – across Australia and New Zealand, op shops (secondhand/charity stores) are piled with cheap sports gear, from surf boards and wet suits to masks, snorkels and mountainbikes, all at bargain prices.

Left top: wild swimming at Mataranka's Thermal Pools in Australia's Northern Territory. **Left bottom:** the New South Wales coast near Tamarama. **Above:** hiking near Cape Naturaliste in Western Australia.

QUEENSLAND COAST, AUSTRALIA

The Queensland coast is one big adventure, but get too carried away with the beach and jungle thrills and spills and it might cost you an arm and a leg. Look for cheaper detours off the main tourist tracks, however, and you'll find thrills without the bills.

OCEANIA

✋ Splash in waterfalls near Cairns

About 14km from Cairns, a series of beautiful waterfalls splash into idyllic, croc-free swimming holes that locals would rather keep to themselves. The Crystal Cascades are accessed by a 1.2km path, and you can walk on to Lake Morris via a steep rainforest walking trail (allow three hours return). *Redlynch Intake Rd, Lamb Range; free.*

✋ Mountain-bike at Atherton

This excellent network of some 55km of trails in the Herberton Range State Forest is easily accessed from Atherton. The free-to-thunder trails are either easy or intermediate, passing through open eucalypt forest, with plenty of steep sections. Atherton Bike Hire (bikehireatherton.com.au; 5 Robert St) will rent you wheels for A$45/55 per half/full day.

✋ Natural waterslides at Josephine Falls

A series of cascades swoosh over eroded granite boulders as the Josephine Creek rushes down from the misty Bellenden Ker Range. At the lower section of falls, water surges over a sloping wedge of rock, creating a natural waterslide into a croc-free splash pool. *Josephine Falls Rd, Wooroonooran National Park; free.*

✋ Cheap snorkelling on the Great Barrier Reef

Reaching the Great Barrier Reef ain't cheap, but at Great Keppel Island the reef comes to you. This serene island is all empty beaches, hearty bush walks over steep, forested hills, and coral gardens that you can snorkel simply by hiking and swimming there (no expensive boat tour required). *freedomfastcats. com; ferry A$45.*

✋ Night-hike through the Daintree Rainforest

Not free, but not expensive and certainly memorable, biologist-guides lead walks through the lush green rainforest of the World Heritage-listed Daintree,

Below: the wonder of snorkelling on the Great Barrier Reef.

shining a light on rare flora and fauna. *daintreerainforest.net.au; tours from $50.*

Discover indigenous rock art
Indigenous people have lived in the area covered by Chillagoe-Mungana Caves National Park for 37,000 years and you can view their creativity at two rock-art galleries, at Mungana and Wullumba, accessible via a short drive and walking trails from dusty Chillagoe township. *parks.des.qld.gov.au/parks/chillagoe-caves; Chillagoe; free.*

Meet koalas in their natural habitat
About 25km southeast of Brisbane, Daisy Hill is an important koala habitat, and the centre here is designed to acquaint visitors with these curious marsupials, with no intrusive photo-posing or koala cuddles. There are also some great picnic and bushwalking spots. *ehp.qld.gov.au/wildlife/daisyhill-centre; Daisy Hill Rd; Brisbane; free.*

Whale watch at Hervey Bay
For most of the year, Hervey Bay is a soporific seaside village – but that all changes in mid-July, when migrating humpback whales cruise into the bay. To see them, book through a small operator like biologist-led Pacific Whale Encounters (pacificwhale.com.au; A$89 per person), or head over to Fraser Island and scan the ocean for free from viewpoints on shore.

Easy wildlife safaris at Noosa National Park
Noosa's unmissable national park delivers spectacular coastal views (expect to see dolphins) and gorgeous beaches like Tea Tree Bay, reached by trails through trees that shelter koalas and other native wildlife. Get here on the accoya-tree boardwalk along the coast from town.

parks.des.qld.gov.au/parks/noosa; Noosa Heads; free.

Explore the Whitsunday Islands
This paradisiacal group of islands is a sprawl of forested hummocks, with pristine beaches edged by teal waters and coral gardens. For a cheap alternative to expensive boat tours from Airlie Beach, head to Shute Harbour, where Salty Dog (saltydog.com.au) offers guided day-trips to South Molle Island with its ruined resort (from A$90), and kayak rental (half/full day $60/90).

Explore Brisbane's busy South Bank
In Brisbane's 17.5-hectare South Bank Parklands, canopied walkways lead through patches of rainforest to performance spaces, lush lawns, restaurants and bars. There are regular free events ranging from fitness classes to film screenings. *visitbrisbane.com.au; Grey St, South Bank, Brisbane; free.*

Walk the Bartle Frere Trail
From Josephine Falls, 75km south of Cairns, this epic free hike makes a 16km loop around Queensland's highest point. This is tropical rainforest country, so expect to work up a sweat before you emerge from the forest for giddying views over the Tablelands to the Great Barrier Reef. *parks.des.qld.gov.au/parks/bartle-frere; Josephine Falls; free.*

Surf and hike Burleigh Heads
The legendary surf breaks at Burleigh Head are sometimes dominated by snooty pros. If you can't catch a break on the surf, head instead to Burleigh Head National Park, where the 1.2km Oceanview Track edges along a 27-hectare rainforest reserve past basalt columns. *parks.des.qld.gov.au/parks/burleigh-head; Burleigh Heads; free.*

ESCAPE THE EAST COAST CRUSH

An easy 30-minute ferry chug from the Brisbane suburb of Cleveland, the unpretentious holiday isle of North Stradbroke Island is the east coast's best-kept secret. There's a string of glorious powdery white beaches and quality, cheap places to camp, sleep and eat. It's also a hotspot for spying dolphins, turtles, manta rays and, between June and November, hundreds of humpback whales. *sealinkseq.com.au; Toondah Harbour, Cleveland;* **ferry one way per vehicle incl passengers from $61.50, walk-on $9.**

OCEANIA

SOUTH ISLAND, NEW ZEALAND

It's easy to find your own cost-effective slice of wilderness on New Zealand's 'mainland'. The only problem: choosing between the sublime forests, mountains, lakes, beaches and fiords that make this island one of the best outdoor destinations on the planet.

✊ Hike the Hooker Valley

The best of Aoraki (Mt Cook) National Park's day walks, this track heads up the Hooker Valley and crosses three swing bridges to the Stocking Stream and the terminus of the Hooker Glacier, beneath Aoraki/Mt Cook. *doc.govt.nz; free.*

✊ Ice and steam at Tekapo Springs

During the colder months (May to September) the Lake Tekapo Springs ice-skating rink and tube slide offers views to the snow-capped mountains, and there are outdoor thermal pools to warm up in after. *tekaposprings.co.nz; 6 Lakeside Dr; ice skating NZ$19, hot pools NZ$27.*

✊ Spot penguins in the Catlins

The Catlins are a beguiling blend of farmland, forest and coastline punctuated by caves, cliffs and blowholes. A tip: to spot penguins at Nugget Point, spoonbills at Pounawea, and sea lions at Surat Bay, all you need is walking shoes and patience. *catlins.org.nz; Otago & Southlands; free.*

✊ Climb Avalanche Peak

It's no Everest but this summit can be bagged in a day by moderately fit hikers. Saddle up with appropriate gear and supplies, check on conditions, and embark on the 1.1km grind to the 1833m-high summit for views to rival the Himalaya.

doc.govt.nz; Arthur's Pass National Park, Canterbury; free.

✊ Westland Tai Poutini National Park

The West Coast's twin glaciers – Franz Josef and Fox – attract big spenders wishing to get close to the icefalls. Hikers, however, can admire them from vantage points all over the containing valleys. *doc.govt.nz; Westland Tai Poutini National Park; free.*

✊ Kiwi-spotting on Stewart Island

There's a good chance of spotting one of Stewart Island's 20,000 tokoeka (kiwis) on pricey night tours. Ask locals about a particular Oban sports field, upon which the bird may forage around dawn and dusk. *Oban, Stewart Island, off South Island; free.*

Below: trekking on Fox Glacier, which can be reached by helicopter or viewed for free from surrounding slopes.

© Niradj | Shutterstock

NORTHERN TASMANIA, AUSTRALIA

Northern Tasmania punches above its weight when it comes to natural beauty, weird wildlife and unspoiled scenery. Luckily for you, obliging nature provides free (or low-cost) encounters with mountains, Tasmanian devils, penguins and more.

Meet penguins at Low Head

Meet charming penguins as they waddle ashore at Low Head Lighthouse, and admire the Kanamaluka/Tamar River spilling into Bass Strait. Inexpensive tours arrive at sunset. *penguintourstasmania.com.au; 485 Low Head Rd, Low Head; per person A$22.*

Hike Cradle Mountain

Cradle Mountain is a spectacular 13km climb with some scrambling near the summit. It's just one of many wonderful trails in a national park home to Tasmanian devils, echidnas and platypus. *parks.tas.gov. au; Cradle Mountain-Lake St Clair National Park; person per day A$25 (with shuttle).*

Traverse 'The Nut'

Known to locals as 'The Nut', this outcrop is all that remains of a 12-million-year-old volcano. Claim the summit after a steep but free 20-minute climb, or a paid ride on the chairlift. Watch for shearwaters at the top. *thenutchairlift.com.au; off Browns Rd, Stanley; chairlift one way/return A$11/17.*

Take an icy dip in a natural gorge

Cataract Gorge brings the wilds into the heart of Launceston. At First Basin you'll find a free (chilly) 50-metre outdoor swimming pool and Victorian-era gardens. *launcestoncataractgorge.com.au; Launceston; free.*

Camp in the wilderness of Flinders Island

There are free Department of Parks campsites all around gorgeous Flinders Island; North East River and Lillies Beach are the pick of the bunch. It's easy to find your own private beach for swimming or combing for topaz shards and nautilus shells. *parks.tas.gov.au; camping free.*

Ben Lomond National Park

In winter, Ben Lomond is a swanky ski destination, but in summer, locals hike the trails. The flower-strewn mountain plateau is stunning in late spring and summer, with gravel tracks for sunny walks – well worth the moderate entry fee. *parks.tas.gov.au; pass vehicle/person per day $24/12.*

Below: whether you climb to the top or skirt the surrounding lakes, Cradle Mountain rocks.

OCEANIA

BEST FREE WALKS

Dawdlers Down Under often have to pay to plod around many popular parks and paths, but here's a selection of freestyle footpaths.

TONGARIRO ALPINE CROSSING, NEW ZEALAND

This fabulous one-day trail tiptoes past active volcanoes and luminous lakes. The weather can be as confronting as the terrain, so be prepared (for almost anything). You'll need to arrange a car drop to get there. *tongarirocrossing. org.nz; Tongariro NP, North Island, New Zealand; 19.5km; free.*

MT FEATHERTOP AND THE RAZORBACK, AUSTRALIA

One of the Victorian Alps' top trails, this two-day mission takes trekkers along a fantastic ridgeline to the second-highest – but first-prettiest – peak in the state. A demanding hike for experienced walkers, it's also cross-country skiable in winter. *parks.vic.gov. au; Vic, Australia; 36km; free.*

CAPE-TO-CAPE, AUSTRALIA

Running between Cape Naturaliste and Cape Leeuwin, this lighthouse-to-lighthouse seven-day epic is one colossal coastal walk, promising everything from whale sightings to wine tasting. Wild camping is possible. *capetocapetrack. com.au; Western Australia; 135km; free.*

REES–DART TRACK, NEW ZEALAND

This challenging four-to five-day adventure near Queenstown follows the spectacular Rees and Dart rivers. Considerate back-country camping is permitted, and free unless you're using hut campgrounds. *doc.govt.nz/parks-and-recreation; Mt Aspiring National Park, South Island, New Zealand; 86km; free.*

ORMISTON GORGE AND POUND, AUSTRALIA

There's more to explore in Australia's Red Centre than the Rock. An offshoot of the epic Larapinta Trail, this dramatic day-long desert adventure takes you through a gorge and the West MacDonnell ranges to a croc-free swimming hole. *nt.gov.au/leisure/parks-reserves; West MacDonnell National Park, NT, Australia; 7.5km; free.*

BARTLE FRERE TRAIL, AUSTRALIA

From Josephine Falls, 75km south of Cairns, hike to Queensland's highest point. Tropical trails climb dramatically through lush rainforest until you pop out of the canopy and catch a view across the Tablelands to the Great Barrier Reef, vaguely visible through the iridescence of the Coral Sea. *parks.des.qld.gov.au/parks/bartle-frere; Tropical North Queensland, Australia; 16km; free.*

MT WELLINGTON AND THE ORGAN PIPES, AUSTRALIA

The best spot to absorb Hobart's end-of-the-world ambience is atop Mt Wellington, the apex of this rock-strewn scramble, which starts with a bargain bus ride into the suburbs. A stern climb earns stunning views over the Organ Pipes (a climbing mecca) to the harbour, where icebreakers prepare for Antarctica. *parks.tas.gov.au; Hobart, Tasmania, Australia; 13km; free.*

BONDI TO COOGEE BEACH, AUSTRALIA

A classic clifftop canter joining up five of Sydney's best beaches, this urban adventure starts amid the backpackers and breakers on Bondi and traces the curvaceous coastline south, taking in Tamarama, Bronte, Clovelly and finally Coogee beaches. Allow ample time for swimming breaks and beachside beers. *bonditocoogeewalk.com.au; Sydney, NSW, Australia; 5.5km; free.*

NORTH ISLAND, NEW ZEALAND

New Zealand's North Island has its own sublime combination of forests, mountains and beaches. Officially designated 'Great Walks' come at a premium, but there are plenty of excellent day-tramps (hikes) where the only cost is sweat and commitment.

🖑 See glow worms at Whangarei

Adventurers duck into this undeveloped network of caverns near Whangarei to see glow worms and the nearby Whangarei Falls. Take a torch, strong shoes, a friend for safety and be prepared to get wet. *Abbey Caves Rd, Whangarei; free.*

🖑 Boogie-board down Rere Falls

About 50km northwest of Gisborne, the Rere River creates a water slide; bring a tyre tube or boogie board to cushion the bumps and slide down the 60m-long rocky run into the pool at the bottom. About 3km downriver, you can walk behind Rere Falls. *Wharekopae Rd, Ngatapa; free.*

🖑 Bathe at Hot Water Beach

At this beautiful Coromandel beach you can dig out your own personal thermal pool in the sand for two hours either side of low tide, then relax in the naturally hot water while enjoying the sociable atmosphere. *Hot Water Beach Rd (off SH25), Coromandel; free.*

🖑 Tongariro Alpine Crossing

One of the finest day-walks in the world traverses the volcanic heart of Tongariro National Park past steaming vents and springs, surreal lakes, craters and ridges offering magnificent views. Arrange a car drop off then walk for free for 19.4km.
tongarirocrossing.org.nz; Tongariro National Park; free.

🖑 Cape Kidnappers gannet colony

One of New Zealand's best bird circuses, a rowdy gaggle of 20,000 gannets gathers on this cliff top. Some visitors come by tractor tour, but you can also walk to them from Clifton (five-hour return). *doc.govt.nz; free.*

🖑 Get your geothermal fix in Rotorua

Thermal parks such as Te Puia are worth the entry fee, but Kuirau Park Precinct, 10 minutes' walk from central Rotorua, has a crater lake, bubbling mud pools and plenty of steam for free. *rotoruanz.com/kuirauparkprecinct; Ranolf & Pukuatua Sts, Rotorua; free.*

Below: many geothermal areas in Rotorua are pay-to-enter, like Pohutu Geyser in Whakarewarewa Thermal Valley, but free alternatives are available.

© Pichugin Dmitry | Shutterstock

OCEANIA

PAPUA NEW GUINEA

If you've come as far as PNG, you've already established your explorer credentials. Seeing the wild interior without joining an expensive tour takes extra dedication and negotiation, but the cultural and natural encounters you'll have are worth it.

✋ Port Moresby Nature Park
This natural bower by the University of Papua New Guinea is an island of calm. More than 2km of walkways thread under and through the jungle canopy, home to exotic plant species and animals. *portmoresbynaturepark.org; Goro Kaeaga Rd, Port Moresby; entry PGK20.*

✋ Hit the waves
PNG's best waves break during the monsoon (from late October to April) along the north coast and in the islands. Bring your own board to Kavieng, the western end of New Ireland, Wewak and Ulingan Bay on the mainland; and Vanimo, near the border with Indonesian Papua. *Free.*

✋ Trek rugged, dormant volcanoes
The dormant volcanoes around eruption-scarred Rabaul are an easy win for trekkers. Start with 688m Kombiu; you'll be up and down in 2½ hours if you're fairly fit and leave early in the morning. *Rabaul, East New Britain Province; free.*

✋ See New Britain's underwater world
Simpson Harbour offers several first-class WWII wreck dives, while the reefs off the western tip of Gazelle Peninsula are a coral wonderland. Kabaira Dive (en-gb.facebook.com/diveandtoursrabaul) offers snorkelling from PGK55 and dives from PGK110.

✋ Cycle the Boluminski Highway
New Ireland's Boluminski Hwy is ideal for cycling, with very little traffic, no pollution, a flat, mostly surfaced road and guesthouses located along the way. You can cover the whole stretch in five days. Arrange bike hire in Kavieng for PGK60-100 per day. *New Ireland Province; free.*

✋ Spot birds of paradise in Tari
The Tari Basin and the Tari Gap are top spots to see colourful birds of paradise. Sir David Attenborough put the place on the map for birdspotters with his 1996 documentary Attenborough in Paradise. There are no fees to enter the forest, but you'll need to arrange transport and a local guide. *Tari, Hela Province; free.*

Above: a volcano smoulders near Rabaul in tropical Papua New Guinea.

OCEANIA

BEST
AQUATIC
ADVENTURES

Oceania is a continent of islands, so water is the main focus for playtime, but some water activities cost fortunes; here are some top aquatic experiences for less.

SURFING AND CAMPING AT CRESCENT HEAD, AUSTRALIA

Surf's up at this chilled-out New South Wales hangout. Test your virtuosity against the right hand, which goes for 400m with good barrelling in the right conditions. Stretching south from the headland, Goolawah Regional Park has gorgeous beach camping from A$24 for two people. *Kempsey, NSW, Australia; free.*

SNORKEL WRECKS AT MORE-TON ISLAND, AUSTRALIA

With five national park campgrounds and easy vehicle and passenger transfers by ferry from Brisbane, Moreton Island has the works. Snorkellers head to the west coast to investigate the Tangalooma Wrecks, 15 in total, sheltering myriad fish and accessible right off the beach. parks. *des.qld.gov.au/parks/ moreton-island; free.*

FRESHWATER SWIMMING IN LAKE MCKENZIE, AUSTRALIA

Cheap forest campgrounds and prime conditions for off-road sand driving attract legions of travellers to Fraser Island, and you can't deny the beauty of the long sand shore and sparkling inshore lakes. With sharks common offshore, swimming in Lake McKenzie has massive appeal, and there's camping nearby for A$6.55 per person. *Fraser Island, Queensland, Australia; free.*

HOT SPRING SWIMMING IN ELSEY NATIONAL PARK, AUSTRALIA

Dyed a surreal, hallucinogenic blue by dissolved limestone, Bitter Springs is a serene palm-fringed thermal pool, 3km from Mataranka along the sealed Martin Rd. The water simmers gently at 34°C in a glade of eucalypts and dry palms, and there's camping nearby for A$6.60 per person. *nt.gov.au/leisure/parks-reserves; Mataranka. NT, Australia; free.*

KAYAK WITH DOLPHINS IN MARLBOROUGH SOUNDS, NEW ZEALAND

A maze of flooded glacial valleys, the Marlborough Sounds provide epic water for kayaking – and there's a good chance of spotting dolphins as you paddle in the deep, calm waters. Rent kayaks or arrange tours with Sea Kayak Adventures. *nzseakayaking.com; Picton, South Island, New Zealand; kayak rental half/full day NZ$40/60.*

PADDLE BY NIGHT IN AKAROA, NEW ZEALAND

Conventional stand-up paddleboarding isn't enough in the Kiwi town of Akaroa. Here, locals paddleboard at night, on boards lit up by LEDs. NightSUP Akaroa offers this unusual activity on the sheltered waters of Akaroa Harbour, or you can roam further afield by day on a conventional board. *nightsupakaroa.co.nz; Rue Lavaud, Akaroa, South Island, New Zealand; night SUP NZ$99, SUP hire per hour from NZ$30.*

GET THE LOCAL BEACH VIBE IN FIJI

Fiji's most famous strips of sand are in the Mamanuca Islands, but getting here involves an expensive day-trip by resort boat. Save your cents by hitting the stunning sand at Natadola, a broad sweeping swerve of brilliance at Sanasana, just off the coast road from Nadi. *Natadola, Viti Levu, Fiji; free.*

WELLNESS

In few places in the world do so many people take so much advantage of the unspoiled nature that surrounds them for physical and emotional wellbeing. From swimming in jungle waterfalls to yoga on the beach before work, the people of Oceania have work-life balance covered, and it's easy to plug in for not much moolah by following their lead.

Things are highly organised in Australia and New Zealand, so there's often a cost, but even paid-for activities come with bonus nature. There are hot pots looking out over pristine valleys, national park

Above: dig your own hot tub on Hot Water Beach in Mercury Bay, New Zealand. Right: dolphins jumping for joy in New Zealand.

© Pete Seaward | Lonely Planet; jacquesvandinteren | Getty Images; Penny Carroll | Lonely Planet; Rolf_52 | Shutterstock